PEOPLE, POLITICS, POLICIES and PLANS

PEOPLE, POLITICS, POLICIES and PLANS

THE CITY PLANNING PROCESS IN CONTEMPORARY BRITAIN

Ted Kitchen

P·C·P
Paul Chapman
Publishing Ltd

Paul Chapman Publishing Ltd
144 Liverpool Road
London
N1 1LA

British Library Cataloguing in Publication Data

Kitchen, Ted
People, politics, policies and plans : the city planning process in contemporary Britain
1. City planning – Great Britain 2. Urban policy – Great Britain
I. Title
307.1'2'16'0941'09049

ISBN 1 85396 359 3

Typeset by Dorwyn Ltd, Rowlands Castle, Hants
Printed and bound in Great Britain

A B C D E F G H 9 8 7

To Ann, Amanda and Christopher

Contents

Preface

There is a very important sense in which I am the source of a lot of the material in this book. I have not hesitated, therefore, to go into the first person to indicate in some instances where this is occurring. The positions that I have held in Manchester and the roles that I have played make me (I would argue) well placed to do this. These have been as follows:

July 1979–July 1983	Assistant City Planning Officer, Policy Division.
July 1983–May 1984	Assistant City Planning Officer, City Centre and Strategy Division.
May 1984–April 1988	Assistant City Planning Officer, Area Planning Division.
April 1988–October 1989	Assistant City Planning Officer, City Centre and Support Services Division.
January 1985–October 1989	Senior Assistant City Planning Officer, working as the City Planning Officer's designated deputy whilst maintaining line management responsibilities as a divisional head.
October 1989–March 1992 and October 1992–December 1993	City Planning Officer.
March 1992–October 1992	Acting Chief Executive; also Acting Clerk, Greater Manchester Passenger Transport Authority.
April 1992–July 1995	Non-Executive Director, Greater Manchester Passenger Transport Executive.

January 1994–October 1995	Director of Planning and Environmental Health. In a very formal sense, as part of an agreed restructure, my title did not change to this until April 1995. The date here, therefore, relates to the date of assumption of functions. The last three months of this period were spent on sabbatical at the Department of Planning and Landscape, University of Manchester, when the first draft of this book was written.

Thus, I was a senior manager in planning and related matters with Manchester City Council for over 16 years. The first decade of this period was spent as senior line manager for every part of the planning service, as the divisional structure changed with reorganisations, The last six years were spent working at chief officer level, and this book mainly draws on that experience.

This book seeks to piece together stories that unfolded as a series of fragments or episodes in which I was an active player, the connections between which were not always obvious at the time. Thus, my perspective was not a disinterested one. My version of events may therefore be a version that other participants would not necessarily sign up for, and since a sizeable element of hindsight has gone into this it is perfectly possible that some of this could be regarded by those other participants as self-justificatory. The purpose of all this, however, is not to provide a chronicle of events as I saw them in the style of memoirs, but rather to try to convey through these experiences an understanding of what the city planning process in a major city is actually like.

I have been much aided in this by reliance not merely on my own memory but on documentary sources. These have included:

- The material that the City Council produces that has acquired the status of public documents, and there is a very large quantity of this.
- Press and other written reports of various kinds, of which there is also a considerable quantity.
- The fact that I kept a daily journal from my appointment as City Planning Officer in October 1989 to my appointment as Acting Chief Executive in March 1992, which not only recorded what I had been doing but which also reflected on what its significance might be. Regrettably, I did not maintain this from that point onwards because of pressure of work, but it is hoped that the more recent period is in any event rather fresher in my memory than this earlier phase of my life as a local government chief officer.
- Parallel literature of all kinds from all sorts of organisations.

Where I have thought it helpful or appropriate, I have cited sources so that my perceptions can be followed up or cross-checked if that is what people wish to do. I have not sought to do this exhaustively, however, because I do not wish to obscure the central point that this book relies heavily on my personal reflections on a period when I was privileged to be given the opportunity to play key roles in the planning and development of one of Britain's most dynamic major cities.

Finally, the usual words about errors of memory or judgement being mine alone need to be written, but it would scarely be possible to find a form of writing about planning where this would be more true than in this case. I am very conscious that, at the end of the day, I am presenting my view in this book, and that I am doing it very selectively, as I must if my objectives are to be achieved. Lots of people are very well placed to say whether or not my perceptions in many of the situations I describe coincide with theirs, or whether my selectivity amounts to distortion. If in particular people whom I have worked with over the years and for whom I have enormous respect feel that some of this doesn't square with what they would have said, then I can only seek their indulgence and ask them to accept this as still further evidence of my fallibility. If members of my former staff, whom I tried to lead and encourage but often in all probability took for granted and disappointed, feel in any way let down by what I have written, then I am genuinely sorry about this; but I do hope they feel when taken in the round that this is recognisably the city planning process in Manchester in which many of them played such a distinguished if unsung part.

Ted Kitchen

Acknowledgements

I would like to thank first the people of Manchester and the staff of its City Planning Department for providing the inspiration that has led to this book. They were a constant source of professional renewal, and I hope they feel that my debt to them has been repaid in some small measure by what has now emerged. May they continue to flourish long after I am gone.

I would also like to thank my colleagues in the Department of Planning and Landscape at the University of Manchester for providing a very supportive environment for me at what could have been a very difficult time during a sabbatical period from August to October 1995, when the first draft of this book was written. My new colleagues in the School of Urban and Regional Studies at Sheffield Hallam University have been equally supportive during the process of turning that first draft into this final product, and they have been both tolerant and understanding of the things that their new planning professor has not been doing whilst this process has been underway.

Patsy Healey, David Whitney and Christopher Wood have encouraged me right from the outset to undertake this project, and they paid the ultimate price by being asked to comment on the second draft, which they willingly did. Their comments were perceptive and helpful, and their understanding of what I was trying to do often appeared to transcend my own. I hope they feel that the effort they put into helping me has been rewarded in the end-product, although of course its failings are despite rather than because of their help and would undoubtedly have been much greater without it. Dick Schneider, of the University of Florida, provided me with useful trans-Atlantic perceptions whilst on sabbatical at the University of Manchester Institute of Science and Technology, and subsequently via the wonders of e-mail, and I hope that he too feels that his contribution to the creation of a British case study about the world of planning practice to go alongside the somewhat greater amount of material of this kind from America has proved to be worth while.

Ann Kitchen transcribed the first and second drafts from my original hand-written script on to our home computer, and Christopher Kitchen helped with his technical know-how to make this possible. Ann not only lived with the ups and downs recorded in the book but, as if that wasn't enough, volunteered to type it even though she knew what my handwriting was like. I have retained the original hand-written manuscript of the first draft for sentimental reasons, and the enormity of what she took on here is obvious. She also chipped in with several very pithy comments which helped greatly with the process of improving the text from the first to the second drafts. My admiration and my gratitude are boundless.

Kevin Mason handled the process of transferring the text between two essentially incompatible systems as if it was just another job, despite my worries at one stage as to whether this was a practical proposition. Sue Hewinson made sense of my rough diagrams, and was quite prepared to experiment with ways of tackling some of the more difficult ones until we found something that worked, in some cases better than I had hoped. Sarah Fidment took on the process of turning my quite substantial amendments to the second draft into a final version with the same speed and cheerfulness that she had tackled every other job she had done for me since my arrival at the School of Urban and Regional Studies. To all these colleagues, and to their colleagues who in turn supported them whilst they were doing this work for me, I extend my grateful thanks.

1

Introduction

The contents of this chapter

This chapter commences with a personal statement of what my perspectives have been in writing this book, and of what I hope to achieve in doing this. It then provides a short section for readers who may not be familiar with the British planning system. It goes on to introduce the structure, organisation and funding of the planning service in Manchester and the city itself, as a context for the material about the operation of the planning process in the city which constitutes the bulk of this book. Finally, it introduces readers to the structure of the remainder of the book.

Perspectives

The best question I was ever asked by a planning student was, 'Is it fun being a city planning officer?' The answer that I gave, that I would still stand by, was, 'The highs are very high but the lows are very low.' I hope that some of the flavour of this comes through in this book. The interesting point about this question, however, was that just about the only way the student could get an answer to something like this was by asking someone like me direct during such a face-to-face opportunity. The reason for this is that, despite what is now virtually 50 years of powerful statutory planning activity in Britain, there is very little accessible material about what the planning job is actually like from the perspective of people who have been able to play a key role in the process. Most of the writing about British planning practice has been done by people who have had little direct experience of it, certainly at a senior level or for a protracted period. Most of the practice of planning in Britain has been undertaken by people who have not written about their experiences and who would probably say that there has been no encouragement or incentive to do so. These diverging paths are described by Hall (1988, pp. 340–41) and by Hoch (in Thomas 1994, pp. 205–21). This is a situation of which the British planning profession should be ashamed, because it certainly means that we have learned far less from operating our planning system than we should have done. In its own way, this book is an attempt to do something about this state of affairs.

The cupboard is not completely bare, however. For example, Wilfred Burns (1967) wrote a short book about his experiences and approaches as City Planning Officer of Newcastle upon Tyne, which is certainly open to the Frank Sinatra charge ('I did it my way') but at least was an attempt to let others understand what was being done there, and why. Cross and Bristow (1983) have also helpfully collected together the structure planning experiences of some senior practitioners mainly in the 1970s, although some of this arguably struggles with being both critical and reflective. Perhaps this is inevitable in a situation where many of the contributors to that volume were still employed by the authorities whose structure planning experiences they were reporting; it can be difficult, and may seem disloyal, to produce reflective material about aspects of one's continuing job. The writing of Urlan Wannop (for example, Smith and Wannop, 1985; Wannop, 1995) also reflects his long experience as a planning practitioner at the regional and subregional scales. There are some pieces by British practitioners about the ethical dilemmas that planning practice can raise in Thomas and Healey (1991), and Thomas has followed this up at a more theoretical level (Thomas, 1994) by collecting together a series of writings about some of the (often neglected) value issues that planning raises. There have also been attempts to provide more vehicles for practice writing in planning, such as Pergamon's *Progress in Planning* series and the emergence of the journal *Planning Practice and Research*. The sparsity of this record, however, does appear to show that as a profession we have ignored the need to ensure that future generations can learn as much as possible from our accumulated practice experience.

There are also some practice-type contributions from people who have been involved in the planning process in various ways. Blowers (1980) has written about the planning process in Bedfordshire from the perspective of someone who is both an academic and a local government politician, and who spent four years in the mid-1970s as Chairman of the Environmental Services Committee of Bedfordshire County Council. More recently, Tim Blackman (1995) has written in part from his first-hand experience as head of research at Newcastle City Council, although his perspective is that of corporate policy rather than planning practice as such. Fragments can also be found in some of the work on evaluating what has actually been done in the name of urban planning (Masser, 1983; Hambleton and Thomas, 1995), although it perhaps reinforces the general point being made here to note that there are few practitioner contributions to these volumes. There are rather more academic studies of these matters, however, which do at least mean that there has been some evaluation of planning achievements and failures. Pride of place should probably go to Peter Hall and colleagues' massive *The Containment of Urban England* (Hall *et al.*, 1973), although this is now over 20 years old. There were also, in the early 1970s, following in the footsteps of seminal US writing by Jacobs (1964) and Gans (1972), a series of very critical and often polemical examinations of what the consequences of British planning practice were turning out to be for people on the receiving end of its ministrations (Davies,

1972; Dennis, 1970; 1972; Palmer's introduction to the British edition of Goodman, 1972; Simmie, 1974), which played a part in the planning profession's radical changes of stance since then, and in the guilt which has been hung like an albatross around the neck of planning in the inner-city since then, to which I refer in Chapters 2 and 7 of this book. One of these (the work of Davies) makes a fascinating contrast with the work of Burns quoted above, since they are radically different perspectives on the same processes in the same city (Newcastle) at the same time.

The 1980s saw the development of reflective and more analytical work, more concerned with trying to understand and to express what planning was and was not achieving than with a polemical viewpoint. Many of these writers worked together for a period at the former Oxford Polytechnic. Prominent here is the work of Simmie (1981) on the planning process in Oxford, Webman (1982) on the process of reviving industrial cities which contrasts Birmingham and Lyon, Healey (1983) and Bruton and Nicholson (1987) on the local planning process, Elson (1986) on conflict mediation in urban fringe and green-belt areas, and Healey *et al.* (1988) who seek to integrate much of this material by looking at the roles of land-use planning in the mediation of urban change. Punter (1990) has also made an important contribution on the achievements of the planning process over a 50-year period in Bristol in seeking to influence the design of office developments. This interest in evaluating what planning is doing is now reflected in the Department of the Environment's research programme, for example in the work of Elson, Walker and Macdonald (1993) in looking at the effectiveness of green belts and of Robson *et al.* (1994) in assessing the impact of urban policy. There are obvious sensitivities around evaluative material of this kind when the policies of governments are themselves being evaluated in government publications. Perhaps this shows how much progress has been made in evaluating planning processes and policies over something like two decades, compared with a situation where the only material of this kind (with the honourable exception of the work of Hall *et al.*) essentially took the form of sniping from the outside.

There is also a plethora of relevant literature from the USA, although one needs to be careful with the direct transplantation of ideas (McCallum, 1976). There is also arguably a larger tradition in the USA of academic writing of case studies, for example from Altshuler's seminal analysis of the city planning process from a political perspective (1969) to a recent attempt at collecting together on a systematic basis experiences of urban revitalisation in several US cities (Wagner, Joder and Mumphrey, 1995). The same problems appear to persist, however, in getting practitioners to write about their experiences. Two exceptions are the books written by Allen Jacobs (1978) based on his experiences as planning director of San Francisco, and by Krumholz and Forester (1990) describing and analysing the experiences of Norman Krumholz and his team when tackling the city planning job in Cleveland mainly in the 1970s from an avowedly equity perspective (in other words, trying to make the planning process work for the benefit of the underprivileged and underinvolved people of the city). Benveniste (1989) also writes challengingly about land-use planning from the perspectives of a policy planner who has been involved both as a practitioner and as an academic. As in the UK, there is far more US material available from academic observers than there is from direct participants.

Quite apart from the learning problems associated with this lack of accessible material (Benveniste, 1989, pp. 110–29; Argyris, 1992), I suspect that planning itself has undersold its solid if unspectacular achievements as a consequence. Krumholz and Forester (1990, pp. 241–42) put this very well:

> We suspect that many other planners have been doing similar but more quiet work: resisting massive projects that threatened the public welfare, calling attention to public opportunities, seeking to inject high quality analysis into political decision processes. Unfortunately, though, many planners, and most of the public at large, never see much of this important work that planners around the country do because their work often involves preventing or resisting public-threatening boondoggles, schemes for private enrichment at public expense, or projects that are just poorly planned and designed. Who can see the monstrosity that was never built? Who knows that these specific millions of tax dollars in a city's budget might have been unnecessarily given away? Who really knows that the transit fare might have been twice what it is, had the planning staff not aggressively intervened? As planners, we can hardly afford to have such invisible work go unrecognized and unappreciated. In the planning profession and the academy too, we need to tell the story of such work, to recognize and indeed honor it, whether it occurs in rural townships or our largest metropolitan areas.

Of course, it isn't just a question of writing up our experiences as frequently or as voluminously as possible. What is important is that this is done in a way that will help others to understand, to begin to learn from, and perhaps to begin looking at some of their own experiences in new ways. Again Krumholz and Forester (*ibid.*, p. 242) put this very well:

> How exactly do we 'learn from experience?' We do not just have experience, we have experience of *something*. 'Experience' is a joke that theory plays on history. Descending from abstract theory to the apparently firm ground of concrete history, we soon learn that experience is endless, seamless, directionless, sometimes eye-catching, sometimes tiresome and irrelevant. Without a sense of direction, we will walk backward rather than forward. Without a broad sense of purpose, our knowledge of historical experience may never seem to matter. Without a sense of pressing questions, we may review our past without ever fashioning answers to the problems facing us today.

In trying to think about these things over a period of my professional career, my guiding light has been the work of Donald Schon. I still regard his *Beyond the Stable State* (Schon, 1971) as the best book about planning yet written, because it was the first time I really felt that someone had helped me to understand the nature of change and of how to cope with it; and this must be the very stuff of planning. In particular, it helped me to see that planning was about much more than the search for the right technical answer. This has led me directly to see planning in the ways in which it is presented in this book. Thus, although it is now nearly a quarter of a century old, that book has been a source of inspiration to me throughout my career despite the fact that it doesn't set out to be explicitly a book about planning. More recently, Schon (1983) has coined the phrase 'the reflective practitioner'. This describes a series of approaches to professional operations which neither rely on claims of technical rationality nor fall on their swords in the face of an onslaught of radical critiques of professional activity, but which draw on both of these whilst staking out approaches to professional work which are relevant and satisfying both for the professional and for those on the receiving end of the professional's ministrations. If I have aspired to be anything in my career as a professional planner, I hope it is this; and I hope too that the approaches of a reflective practitioner do surface in this book. In particular, I would hope that it would be seen as a contribution to what Schon describes as 'repertoire-building research', one of the four types of reflective research he describes (*ibid.*, pp. 315–17).

I recognise that my own attitudes and values are a component in the stories which I tell in this book, and this is not something that I would wish to hide. Allen Jacobs, when reflecting on his experiences in San Francisco, describes this in a manner that I would be happy to claim as my own (Jacobs, 1978, in Stein, 1995, p. 434):

> City planners should not be neutral, and I do not believe their clients, at the level of local government, expect them to be without values or opinions. After they have arrived at some position, some point of view, some desired direction, one would hope to see it reflected in both public plans and day-to-day recommendations. Why hide it? Further, city planners should be willing to stand up for their points of view if they want to be effective. They should be prepared to 'mix it up'. They must do more than recommend. Within a democratic process they should advocate and search for ways to carry out their plans. I believe, too, that they should value and nurture their utopian predilections. They are nothing to be ashamed of. I do not believe we have done these things enough. We have tended to be meek.
>
> I am not suggesting for one minute that city planners do battle with every person with whom they disagree or with every interest that is different than theirs. Nor am I suggesting that *every* matter that comes up has a right and a wrong side for the planners. Some matters have reasonable alternatives, not just one answer. Some will come out all right no matter what point of view prevails. In any event, the planners must do an honest job of evaluating various courses of action, and they must make their evaluations public. However, I am suggesting that there are many matters that do have right and wrong sides for planners and that when this is the case they should be prepared for conflicts, even with those interests they might wish most to serve.

All this, then, is where I am coming from. As I moved from being a local government chief officer in a major city to a new life in the academic world, I wanted to reflect on my experiences both because the story ought to be told and because my criticism about the lack of literature of this type is not likely to carry weight if I failed to take the opportunity to do what I am urging on others. By telling the stories of the planning process in Manchester during a period of the city's history when its national and indeed international profile as a city regenerating itself was high, I hope I will help both students and existing practitioners to think about what the process of planning is actually like in the real world, dealing not only with the major projects and initiatives that receive extensive media attention but also with the day-to-day job of trying to improve the welfare of our citizens through our powers and the opportunities presented by our position as planners. If the process of telling the stories in this way helps people to understand not only what happened but also how and why, and in particular what judgements or actions were necessary to make things happen, then it will have been a worthwhile exercise for this alone.[1] If, in addition, the material presented enables people to look at the planning process from more theoretical perspectives by providing a basis for grounding their interests and concerns in empirical observations (Glaser and Strauss, 1968), then as far as I am concerned that will be a useful bonus; but it is not my primary concern.

In all probability, the ways in which people, politics, policies and plans interact in a particular location are many and various and perhaps even unique. Thus, a book which is about these interactions in Manchester in the late 1980s and early 1990s is dealing at least in their detail with specific phenomena. This is, of course, one of the well-known difficulties of seeking to generalise from an individual case study. But I would argue that at a broad level there are many features of the processes described in the following pages covering both the way the planning system operates, and the policy issues it tackles, which my colleagues working in local planning authorities throughout Britain, and quite possibly in other countries as well, would recognise. I would also argue that we will never effectively understand how people, politics, policies and plans interact, unless case studies which attempt to be as open as possible about these matters are written, despite the fact that each one may well contain idiosyncratic elements. What follows is offered unambiguously as a Manchester study, which is of interest in its own right, and of value in showing how the planning process actually works in contemporary Britain.

I have tried to do this in a way which focuses on the process rather than on the detail; on the picture rather than on an individual; on the broad thrust and the planning contribution to it, rather than on the daily and weekly ups and downs. I hope this conveys a flavour of what the planning process has been like in a major British city over a number of years from the perspective of someone who played a number of key roles in it over that period. I hope also that my reflections, conclusions and comments will contribute to a whole series of continuing debates, both locally and more widely, about the nature of the planning job in our cities and the changing world to which it has to relate. This book, after all, is only a reflection on a particular period which had no actual boundaries to it and no guarantee of neat endings to all my stories. Indeed, this is apparent on several occasions, where the phrase 'at the time of writing' precedes a statement about an as-yet unknown outcome; this usually means that the outcome was unknown to me by the end of summer 1996. To bring some of the main stories up to date, I have added a short Postscript (p. 224). If this book achieves these objectives, I will feel that it has been worth while.

I had perhaps better say something as a consequence about what this book isn't:

- It isn't a detailed case study, itemising exactly what happened on this day or that, but is rather a synoptic overview.
- It isn't a serial description of the life of a city planning officer. I have deliberately avoided a chronological approach in the structure of the book in order better to achieve the objectives that I have set.
- It isn't an exposé of the inner workings of Manchester Town Hall. Whilst there might have been some very local interest in some of this, it would have got in the way of what I was trying to achieve.
- I hope it isn't an exercise in self-glorification and self-justification. I hope I have avoided this at least to an extent by writing a book that is not about me, but is about a play in which I was a player.
- I hope it isn't a book which tries to present the planning service in the best possible light by only choosing to tell those stories that reflect well on it. I have had to be highly selective in my approach, but I have chosen to tell stories which show the Planning Department in political difficulties (as it undoubtedly was at times) as well as stories which are about successes.
- I have tried to ensure that it isn't a book about Manchester which merely repeats some of the 'civic boosterism' which is an inevitable part of the job of a senior local government officer in trying to promote the welfare of the city he serves. At the same time, I hope my feelings for the city and for the overwhelming majority of its people shine through, as does my respect for its history and for the need to understand this if we are to handle today competently.

I think I can demonstrate that these views about the importance of the city's planning history being understood as a backcloth to contemporary decision-making are not just inventions for the purpose of this book. I was able, for example, to initiate the processes which led to local photographer Len Grant being commissioned to photograph major change processes as they were actually happening in the construction of the new concert hall at Lower Mosley Street and the construction of the new arena at Victoria Station, so that on at least these occasions we wouldn't be as careless with our contemporary history as too often we are. Len's books about the new arena (Grant, 1995) and the new concert hall (Grant, 1996) demonstrate effectively how much we miss by not doing things like this. More recently still, I was able to persuade the City Council that it should regard 1995 as the 50th anniversary of postwar planning in Manchester, since 1945 saw the appearance of the Nicholas plan for the postwar reconstruction of the city and an associated major exhibition. As a consequence it initiated a series of events (an exhibition, a book, the inauguration of a planning archive, the creation of video material for educational purposes) that tell the story of this important and continuing phase in the life of the city in accessible ways. Whilst there was an element of invention of an anniversary to provide an opportunity in all of this, I felt very strongly that the stories ought to be told locally in ways that might help people understand contemporary actions a little more clearly. I certainly feel that the book that this initiative produced (Manchester City Council, 1995d) achieves this objective and could be a model that other cities might adapt to their own circumstances. It doesn't deal with policy conflicts both resolved and unresolved, as 'official' histories will tend not to do when the sensitivities of authors as employees are remembered; but it does draw together a great deal of previously disparate material about the city's planning history, something that had not previously been undertaken.

The British planning system

This book is about the operation of the British planning system in practice in a major city. It makes a series of assumptions about readers' knowledge of the basic characteristics of that system. Readers who are familiar with the British planning system will find no difficulty with this, and can therefore skip this section of this chapter. For readers who are not familiar with the essential features of the British system, this section attempts to describe it in simple terms. Fuller introductions to the British planning system are provided by Greed (1993) and by Rydin (1993) and to the British local government system within which planning operates by Wilson and Game (1994) and by Fenwick (1995). Much of this material is usefully integrated in Cullingworth and Nadin (1994).

At its heart, the British planning system operates through local councils, although there are also overseeing, guiding and in some cases deciding roles for central government. The basic principles of this system were established by the Town and Country Planning Act 1947, and although there have been changes to some of the details of this system, in its essentials it has survived for virtually 50 years. Local councils acting as local planning authorities can either be part of a single-tier system, as in the metropolitan areas, so that Manchester City Council operates as an all-purposes planning authority, or they can be part of a two-tier system as in some of the shire county areas, where functions are split between an upper-tier county council with strategic planning responsibilities and a lower-tier district council with local planning responsibilities. This system operated in the metropolitan areas between 1974 and 1986, so that between those two dates, a Greater Manchester County Council existed. This had strategic planning responsibilities for the whole conurbation within which framework Manchester City Council sat as a district council. I refer to this period on several occasions throughout this book. Apart from this 12-year period, however, Manchester City Council has operated as an all-purposes local planning authority since the 1947 Act came into force.

It is important to note in this context that it is the City Council that is the local planning authority; not its Planning Committee as such, and not its City

Planning Officer as such. For the efficient discharge of its planning functions, the City Council, like most local planning authorities, chooses to establish a planning committee or equivalent, and to delegate most of its planning powers to it. The nature and the degree of delegation have both varied; but this is a matter of choice by the City Council for good practical reasons. Similarly, the City Council appoints planning staff headed by a city planning officer, but this too is a matter of the exercise of local discretion. The City Planning Officer, as a local government chief officer, performs the three classic functions of *adviser* to the council in respect of its planning functions, *implementer* of the council's decisions as local planning authority, and *manager* of the human, financial and other resources that the City Council chooses to devote to the discharge of its planning functions. This was my job for several years in Manchester, and it is the experiences accrued in performing these tasks on which I am primarily reflecting in this book. But it is important to remember that a city planning officer, in carrying out these tasks, is acting as the servant of the city council in which the functions formally reside, and is doing so in a manner which is regulated by a planning committee which consists of elected members of the city council (in other words, elected local politicians) and which formally takes many of the most important planning decisions on the council's behalf. This is why one of the most interesting features of the British planning system is the relationships it sets up between elected local politicians and paid professional officials, and I comment on these relationships on many occasions in this book.

Planning powers given to local planning authorities by Act of Parliament consist very broadly of three types:

- The power to make a development plan to guide the future physical development of the city within a framework provided by the government, which once approved or adopted has elements of legal force behind it.
- The power to require and then to determine applications for planning permission, which, for all but the smallest physical developments, have to be made to the council. These decisions are subject to a right of appeal by applicants to the government (in practice, very largely to an agency called the Planning Inspectorate specifically set up by the government for this purpose) against refusals or against conditions imposed on approvals. With these development control responsibilities also comes the power to take enforcement action against breaches of planning controls.
- A series of environmental powers, which deal with such matters as the ability to carry out works of environmental improvement, the ability to declare areas of particular character or quality as conservation areas, within which special powers designed to retain and enhance that character then apply, and the ability to preserve trees or groups of trees which contribute especially to environmental quality.

The complexity of the legal and administrative processes associated with this range of powers has grown enormously over the 50 years that the postwar planning system has been in operation. Much of this book is about how these powers get discharged, and about how issues get resolved in the context of these powers, and the process of telling a series of stories about these activities is also a process of introducing some of this operational complexity.

An over-riding point which should be made about the British planning system if its practical operation in Manchester is to be understood by readers who are not familiar with that system, is its *discretionary* nature. The British planning system does not operate by producing a development plan and then by looking up answers in it and slavishly following them. The development control system, for example, has a strong component in it of looking at a case on its merits, within which the relationship with the content of the development plan is an important consideration, but not the only one. Similarly, there is a great deal of discretion to be exercised about the ways in which powers are, or are not discharged. Typically, the local planning authority makes a judgement in all the circumstances, in which local knowledge is often a more important component than any other, and where development plan policies give a series of useful starting points and leads, but don't of themselves make the decision. One of the reasons why elected members have such an important part to play in the British planning system is that this widespread process of the exercise of discretion is very often ultimately a political act; it is about what values or aspirations or interests to emphasise, acting on behalf of the community as a whole. The powers of local planning authorities are circumscribed in many ways, but fundamentally they exercise discretion by trying to do what they think is appropriate in all the circumstances and within the range of the planning powers at their disposal. This book attempts to illustrate how these processes of exercising discretion operate in practice.

Finally, and precisely because it is the council that acts as local planning authority as well as local authority for many other functions and services, *planning powers do not operate in isolation*. Councils have many aspirations for the places and for the people for which they are responsible, and planning powers are amongst the tools available to them to work towards achieving their ends in these terms. Although the law properly prescribes certain situations where planning powers must be seen to be operating independently of other powers (for example, to prevent councils favouring their own property-owning interests at the expense of other property owners with equal rights), none the less planning powers are exercised by councils as a relatively small component of a large organisation operating with a relatively large budget across a very broad span of activity areas. Planning powers must therefore be seen not only in their own right but also as part of the corporate operations of a body charged with very wide-ranging responsibilities for the welfare of its area and its people. There can be tensions between these elements, as this book will illustrate, but this does not detract from the point that the planning service is one local government service amongst many, operating as part of this whole.

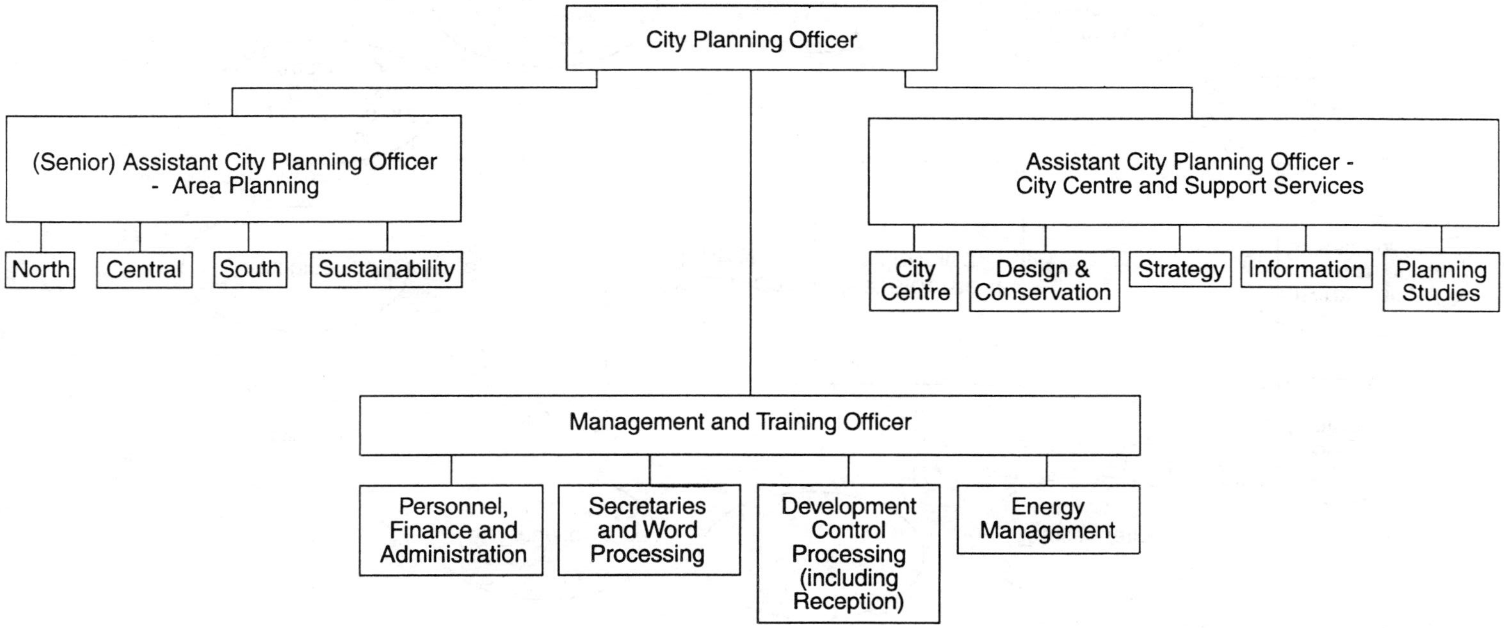

Figure 1.1 Group and senior management structure of the City Planning Department at 31 December 1993

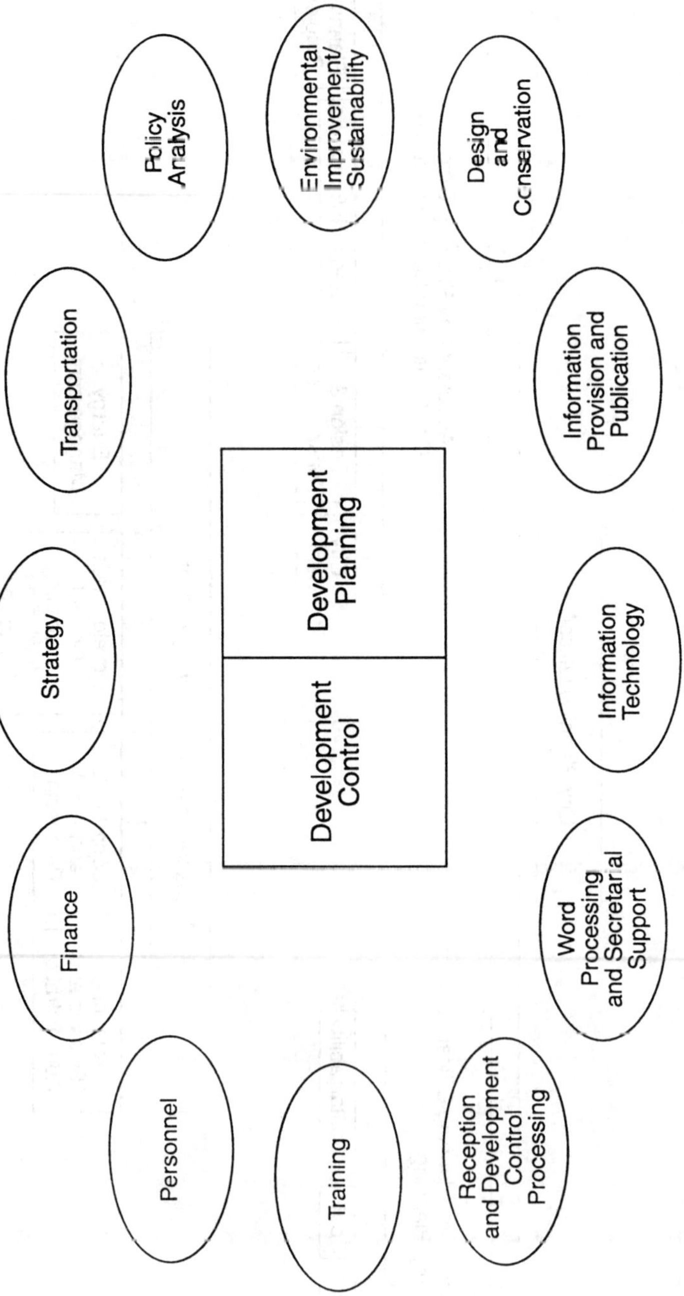

Figure 1.2 Support services in relation to core functions

The structure, organisation and funding of the planning service in Manchester

Until the amalgamation of the Planning Department with the biggest part of the previously separate Environmental Health and Consumer Protection Department in 1994/95, the basic structure of the department had not changed very much for several years, although some changes had occurred from time to time. The group and senior management structure as it was just before the 1994/95 amalgamation is shown in Figure 1.1.

The basic organising principle behind this structure was that the department should consist of a series of most-purpose area teams, supported and reinforced by another series of specialist groups. The reasoning behind this, which would probably not be seen as a typical form of organisation for a large city planning department, is that it is a structure designed to enable the planning department to relate as effectively as possible to its customers, who are discussed in more detail in Chapter 2. The two professional planning divisions each consisted of a mix of area teams and of specialist services, although the balance varied between the two. As its name implies, the Area Planning Division consisted mainly of area teams, whereas the City Centre and Support Services Division linked the work of the City Centre Area Team to a series of strategic and policy support functions with which city-centre work is strongly connected. This mixed structure had the useful by-product of emphasising that area teams and specialist groups were part of what ought to be a mutually reinforcing organisation. An attempt to create one area teams division and one specialist groups division might have emphasised a divide in these terms; and from time to time there were enough difficulties of this kind. The links between these Divisions and the Management Division were also important, because the provision of a professional planning service depends on the support provided by the staff of the Management Division in discharging functions such as finance, personnel, secretarial and word-processing support and reception. Figure 1.2 shows the range of functions delivered by the department, clustered around its two core statutory activities of development planning and development control which are primarily carried out by the area teams. There isn't the space here to go into the nature of each of these functions, however. This approach placed a large management burden on the two Assistant City Planning Officers (one of whom was designated 'Senior' in order that he also acted as my deputy) in terms of the achievement of these kinds of integration, and also in terms of the development of linkages with corporate and political processes, and in a different kind of way on the group leaders as the main focus for the delivery of day-to-day services. With myself, the two Assistant City Planning Officers and the Management and Training Officer made up the Departmental Management Team, which met weekly.

The staff resource position of the Planning Department declined fairly spectacularly over the period of the late 1980s and the early 1990s. The establishment that was agreed for the department on the abolition of the

Table 1.1 Manchester City Council: staffing position (Non-manual full-time employees) at March 1994 in comparison with averages for all metropolitan districts

Service	Staff per 1,000 population Manchester	Staff per 1,000 population all metropolitan districts	Comparison (all metropolitan districts = 100)
Engineering and construction	2.70	1.14	237
Libraries, museums and arts	0.99	0.47	211
Architectural	0.55	0.27	204
Housing	2.38	1.27	187
Other education staffing	5.03	2.71	186
Recreation, parks and pools	0.85	0.56	152
Other services	0.77	0.52	148
All Manchester City Council services	29.98	21.02	143
Central services	2.89	2.07	140
Personal social services	3.84	2.81	137
Estates	0.17	0.13	131
Environmental health	0.48	0.37	130
Lecturers and teachers	9.07	8.36	108
Planning	0.27	0.34	79

Source: Audit Commission, 1995, p. 5. The basic information is stated there as having been obtained from the Local Government Management Board.

Greater Manchester County Council in 1986, when Manchester City Council again became an all-purpose local planning authority, was 179 posts, although this was never quite achieved. By 1993/94, because of progressive budgetary reductions in large or in small steps, the number of staff that the department could afford to pay was in the range 110–15. Although the establishment was a little larger than this, the exact number of staff at any time was a function of what posts we chose to fill at what levels within the cash-limited budget that was available to us. What this actually meant both in relative and in absolute terms was that Manchester, over a period of less that a decade, had shrunk from being a relatively well endowed planning service in staffing terms to one that was poorly resourced. The comparative situation both with other City Council services and with planning services in all metropolitan districts as at March 1994 is shown by Table 1.1.

This table, which puts services in rank order by reference to the comparison between Manchester City Council and all metropolitan districts, shows that the planning service was the only City Council service operating at below the average for all metropolitan districts (at 79 per cent of the average position), and that it operated at little more than half of the average position for all City Council services. Whatever else this table demonstrates, and there are difficulties associated with these kinds of data, it shows a planning service that in terms of resources had not been well supported by the council's political leadership for quite some time.

Underlying all this, of course, was the relative deterioration in the City Council's overall budgetary position during the late 1980s/early 1990s, primarily as a result of ever-tightening central government controls. Whilst Table 1.1 suggests that there was an argument to be made about the relative treatment of the planning service (and it was!), the service was actually very small beer when looking at the City Council's revenue budget as a whole. Very broadly, the planning service for 1994/95 was responsible for less than 1 per cent of the council's revenue budget, and so decisions about the overall budget were usually very little influenced by decisions about planning. However, regimes introduced as part of overall budgetary control had also to be applied to the management of the planning service, and in practice this tended to mean three things:

- Income was counted against the overall budget for the department to reduce the direct burden on ratepayers/local tax-payers, and a great deal of pressure was put on the department to generate income, so that by 1994/95 this funded approximately one-third of net revenue expenditure. The biggest item in this was fees for planning applications. This represented a very major change from a historic position where income had been treated as a receipt for the centre rather than as something which counted against the direct expenditure of the department.
- Expenditure had to be carefully controlled to keep within budget, and if income targets were not going to be achieved (for example, when the recession in the early 1990s adversely affected the number and the size of planning applications, and thus the fee income they generated) the difference had to be made up from savings elsewhere. A particular problem of this kind was looming for the 1996/97 financial year because of the disappearance of Central Manchester Development Corporation (CMDC) and hence the loss of the CMDC development control agency fee, since mainstream revenue expenditure had already been reduced in the late 1980s when this income was substituted for it. Management decisions thus always had to be taken with a close eye on the budget.
- Interdepartmental transactions had to be perceived, monitored and controlled in budgetary terms, although in some cases, of course, these were at zero net cost to the authority as a whole because they were merely transfer payments. This will grow as compulsory competitive tendering spreads throughout local government, and as an internal trading market develops.

This, then, was the resource management context within which the planning service was operating by the early/mid-1990s, and it was a much more restrictive one than had existed in previous decades. One of the things that Brian Parnell, my predecessor, said to me when we were talking about this change subsequently was that for most of his time as City Planning Officer about 20 minutes per annum had been spent on budgetary issues. He was, of course, exaggerating to make a point; but by the early/mid-1990s the City Planning Officer was expected to operate as the manager of a medium-sized business with a turnover of several million pounds per annum and a rigid expenditure ceiling. This had certainly not been the case in previous generations.

The process of achieving the combination of the previously separate Planning and Environmental Health and Consumer Protection Departments in 1994/95 is one of many stories not told in this book, but this is not to minimise its significance. It changed the scale of the operation by more than doubling its staffing, by doubling its revenue and capital budgets when taken together, and by adding about 50 per cent to the department's annual income; so it clearly was a major matter in terms of the management job it entailed. In particular it also changed the way in which the various kinds of support services were viewed, since some of these were common to the two previously separate departments and others offered the opportunity to look at the case for creating new linkages, by putting components of the two together. And in a broader sense, it offered the potential to explore what gains could actually be achieved in terms of standards of service to customers as a result of this combination. At the same time, it is possible to exaggerate the effects of a change like this on the day-to-day jobs carried out by members of staff, since essentially most of them continued to do what they had previously been doing. Changes such as this are a frequent occurrence in local government life, but it is possible to look at them in different ways. Dunleavy (1991) talks about this in terms of 'bureau-shaping' processes, which are about changing how an organisation is structured in order to relate to its environment, and about 'budget-maximising' processes, which are about the distribution of resources within the organisation and sometimes also its ability to obtain resources from elsewhere. In these terms, this particular piece of restructuring in Manchester was a bureau-shaping exercise taking place within a series of assumptions about fixed budget levels.

The City of Manchester

The administrative City of Manchester is the core city of the Greater Manchester conurbation in northwest England, an urban agglomeration of 10 metropolitan districts each with its own unitary council responsible for the provision of most local government services. Figure 1.3 shows the basic structure of the Greater Manchester area.

Manchester is the seat of much regional government, of the largest retail, commercial, office and higher education concentrations in northern England, the location of a great deal of transportation infrastructure, and the home of a large number of cultural, media and sporting activities of regional and national significance. These are the things for which the city is mainly known, but they are provided in and by a city with a 1991 resident population (including students) of just over 435,000, as compared with a population for Greater Manchester as a whole of approximately 2,500,000. The advantages and the problems of being a major regional centre with a relatively small domestic population base are an important contextual element in many of the stories told in this book.

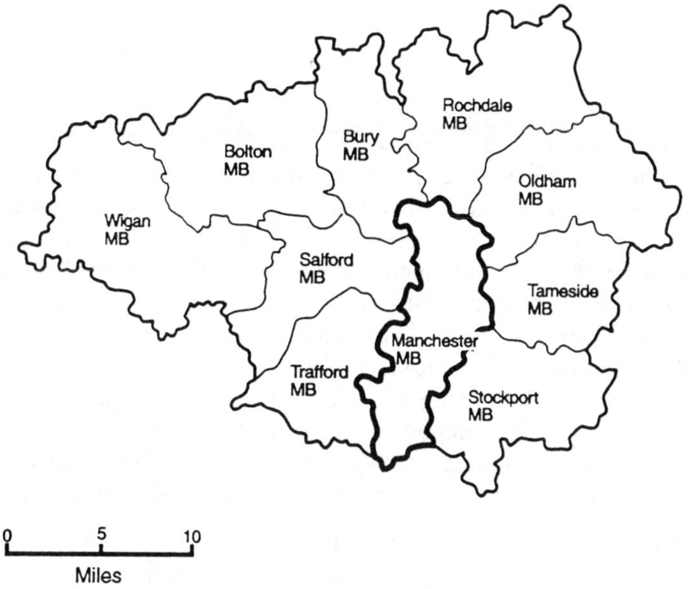

Figure 1.3 Administrative Districts in Greater Manchester

Alongside this range of higher-order activities, Manchester has a large concentration of what are usually termed 'inner-city problems'. The tensions that arise between the coexistence of these two are another major strand in this book. Chapter 7 deals in more detail with inner-city issues specifically, but they will be referred to throughout.

A third strand that constantly surfaces is the city's history and the contemporary legacies to which this has given rise. As in many ways the first city of the Industrial Revolution (Briggs, 1982), Manchester has a name in the world quite out of proportion to its current size. But this also brings with it attitudes and attributes which are a very important part of the contemporary planning job in Manchester, and this is also referred to on many occasions.

The structure of this book

This is not a chronological story of planning in Manchester as I have experienced it. Not only would this be a very difficult thing for an author to do in ways that convey any real understanding of the processes at work but it would also be very confusing for a reader because of the large number of events or stages in those processes that happen in parallel or in multiply overlapping proliferation. What I have therefore chosen to do, in the belief that this will enable me to describe more effectively what the city planning process was like, is to divide the book into three clusters of chapters following this introductory chapter.

Part I deals at some length with what I regard as the main actors, having introduced the planners themselves in this chapter – the customers of the planning service, and the elected members who in a democracy exercise supervisory control over it on behalf of the council.

Part II looks at the main tools available to the planner in setting about the task of helping to improve the quality of people's lives by seeking to improve the places and the environment within which they lead their lives; the development plan-making process, and the development control process.

Part III looks at what have been some of the major arenas within which in recent years a significant planning contribution to developments in Manchester has been made. It has been necessary to be particularly selective in this part of the book, since there are a very large number of potential areas that could have been looked at in these terms. In my inaugural lecture as Honorary Visiting Professor in the Department of Planning and Landscape at the University of Manchester (Kitchen, 1996b), which was delivered while I was still the City Council's Director of Planning and Environmental Health, I identified four issues as being of outstanding importance for Manchester:

- Improving the economic base of the city.
- Tackling more effectively the problems of the most deprived sectors of its population, particularly in the inner city.
- Achieving an effective transportation system.
- Making the city a more sustainable place.

Part III therefore contains chapters on the city's economy, the inner city, transportation planning and sustainability and Local Agenda 21.

In addition, there is a single concluding chapter which returns to some of the ideas introduced in this chapter about the purposes of this book, and which also tries to look forward by reflecting both on Manchester's experiences and on the light these may throw on some of the issues which will need to be tackled in seeking to secure appropriate futures for Britain's major cities.

Note

1 See Forester, in Fischer and Forester, 1993, pp. 186–209, for a very clear exposition of the reasons for attempting something like this.

PART I

THE MAIN ACTORS

2

The customers of the planning service

Background and context

It would not have been regarded a few years ago in British planning circles as an orthodox start to a book about planning practice to begin it with a chapter about the customers of the service. Indeed, at least some of the leaders of the profession would have argued strongly that planning's responsibility for and to the general public's interest would have precluded such a more segmented approach (for some of this debate, see Hague and McCourt, 1974; Simmie, 1974). On this basis, almost by definition, the comprehensive and rational planner had a wider view than any individual interest could possibly encompass, and by a process of elision it was not difficult to see how this could easily slip into the perception that such a view must also be somehow more worthy. It is not too difficult to attack a view of this kind, and the following arguments at least spring readily to mind from my own experiences:

- It is undoubtedly self-serving, and can be self-fulfilling.
- The test of time does not necessarily show that planning decisions stemming from this view of the world are inherently better than anything else that might have been achieved.
- Many people have shown through their lives and through their actions that they did not regard planning decisions as being in their interests, including in inner cities where whole communities were on the receiving end of the redevelopment process (see, for example, Goodman, 1972).
- It is an approach which does not promote both consultation and listening as genuine attempts to find out what people want, and it can easily turn these processes into vehicles for *post hoc* rationalisation (see Davies, 1972; Dennis, 1970; 1972).
- It is at risk of turning its back on the value of the knowledge and awareness base of local people and interests, especially where it promotes a 'we know best' attitude.

There are undoubtedly also more philosophical worries that can be added to this list of essentially experiential arguments, about the elusive nature of the public interest as a concept in its own right (see, for example, Schubert, 1960). Whilst this produces in combination a powerful critique of conventional thinking about comprehensive planning and the public interest, it does not deal well with three rather uncomfortable propositions:

- That this model, whatever its limitations, has not yet been replaced by wholly satisfactory alternatives. Benveniste (1989, pp. 56–86), for example, talks about six theories of planning, of which 'the comprehensive rational approach' is the first, and every one of which he analyses in terms of myths and realities. Taylor (in Thomas, 1994, pp. 87–115) argues that the concept of the public interest is of some continuing value and that the difficulties that have been experienced with it are in substantial part because of the ways it has been used, often to try to justify preconceptions or élite views.
- That there is quite a strong sense in British planning law and practice that this approach is enshrined in the Town and Country Planning Acts and in the decision-making relationships that exist on planning matters between local planning authorities and central government. The disputes in these terms tend not to be about theoretical difficulties over who (if anybody) is entitled to operate in the public interest, but rather about whether in practice particular holders of this torch have got it right in particular cases.
- That there is a sense, at any rate in British planning practice, that the political process which ultimately actually takes planning decisions operates as a kind of 'deemed public interest'. This argument in effect says that elected representatives are there to operate in the public interest, that they have been mandated to do so via the electoral process, and that as a consequence their actions represent an expression of the public interest at any point in time as it is translated into practice via the public decision-making processes that legitimately exist. Whatever political scientists may think of this, I can say from experience that this is broadly how a lot of politicians actually see themselves as operating, even if it may not be expressed in quite such language or always even understood in quite such a way.

Thus, this debate is not about replacing an old wrong with a new right. Not only is it more subtle than there merely being two polarised alternatives but also it needs to be seen in the context of the practicalities of a complex and evolving political and administrative world. Much of my own thinking about the need to be explicit about who the customers of the planning service are, and about the value in practice of doing this, has been heavily coloured by these kinds of considerations. This is one of the reasons why I described Schon (1971) in Chapter 1 as being for me such an important book. In recent years also, it has undoubtedly been affected by the philosophies of the Thatcher governments of the 1980s and the things they did to the planning service as a consequence (Thornley, 1991), whatever one thinks about the generality of this. In particular, the emphasis on value for money against a shrinking resource base, on the need to see the planning service as contributing something of worth and not merely existing as a series of bureaucratic hurdles, and on the need to develop explicit indicators of performance, ought to force practitioners to think hard about what they are doing, for and to whom and why.

The basic conclusions that I have drawn from all this are as follows:

- That the planning service should see itself as having a series of customers, with differing needs, aspirations and methods of working.
- That the planning service should also see itself as having a responsibility to develop, to articulate and to test against customers' perceptions and understandings a series of visions and objectives for the future development of the place it serves, which provide a context for decision-making which planners should constantly be prepared to adapt in the light of changing circumstances and changing customer feedback.
- That the achievement of as much customer satisfaction as is possible as a result of the operation of the planning service is an inherently desirable objective.
- That conflicts between individual customer satisfactions, and between these and visions and objectives for the future of our places, should ultimately be resolved through the political process. It is also for the political process to determine who should legitimately be regarded as the customers of the service, in so far as there may be any doubt about this. This means that the formal decision-making processes of the council employing the local government planner have a special place amongst this range of customers.
- That planners need to understand the sets of relationships that all this gives rise to, the powers and the limitations inherent in their positions both as brokers in this situation and as servants of it, and their responsibility as professionals to give accurate, helpful and honest advice to the participants in this process.
- That the tensions and difficulties that this creates for the position of the planner need to be understood and faced.

There is a particular Manchester dimension to all this, which I suspect would be replicated in other large cities, which has been an important component in the development of my thinking. The inner-city redevelopment process that proceeded in the city from roughly the mid-1950s to roughly the mid-1970s did so by using the planning process, for example through the promotion of compulsory purchase orders, and in some part also by being driven by it. From this perspective, whether or not the planning service actually supported each individual manifestation of that particular phase of inner-city redevelopment is less important than the fact that this was often represented as and was often perceived as the product of planning. This has created a legacy in the inner city of hostility and suspicion towards the outputs of planning which is frequently fanned by the media; how often in all the media do we see development issues presented on the basis that sensible decision-making and the output of the planning service are two mutually exclusive concepts? It is also a legacy that is used by the political process for its own purposes. Labour manifestos in Manchester in the 1980s and 1990s, for example, continued to 'look forward to' the opportunity to redevelop past 'planning failures'. There is a debate to be had, for example, about how fair it is to blame planning for the high-rise system-built housing of the late 1960s/early 1970s (Dunleavy, 1981), much of which across the country as a whole has now been demolished. There is also a debate to be had about how much planning is to blame for the high-capacity urban roads which carved up the inner city to reduce by a few minutes driving times from suburb to city centre. But what is beyond debate is that planning is and continues to be blamed for these things, whatever the rights and wrongs of this. And what is also beyond debate is that planning and planners did participate in these processes and cannot simply wash their hands of them as a consequence. I always felt that my generation of planners had to accept this stigma for what it was, and had to try to show by what they did and the ways in which they did it that planners were not in the business of repeating history but were in the business of trying to add value to the quality of people's lives. This inevitably meant approaching the task in a way that was much more people driven than before, because it sought to include people in the process, and thinking about people as the customers of the process is but a short step from this position.

There is also a cautionary dimension to this. Planning tends to deal with artefacts that have quite a long lifespan to them, and this of itself ought to imply some warnings about deciding in haste and repenting at leisure. In addition, attitudes and values change within and between generations, so that (for example) what was accepted behaviour by public organisations in the 1960s would be unacceptable in the 1990s even though many of the products of that behaviour will not be half-way through their anticipated lives by the time this change has taken place. This too ought to be a potent source of warning about what we are doing today. It also needs to be recognised that the things that people tend to love about cities are the things that are familiar to them, especially when that familiarity goes back a few generations, and that people are often resistant to change if they can see no good or necessary reason for that change. Very often, what has survived from the past represents its best rather than its worst, and the rosy glow of public support for Georgian and more recently Victorian architecture conveniently forgets that these periods also spawned many very poor buildings which have not survived. So why should the things we do today be any different? What likelihood is there that public support for and ultimately love of redevelopment projects will be generated, when complete areas of cities are cleared and rebuilt all at one point in time with little or no effective involvement in these processes on the part either of the people to be moved out or the people to be moved in, especially when some of these governmental processes were neither very open with the people affected by their actions nor very good at dealing with some of the individual problems to which these actions gave rise? Yet this is exactly what did happen in the slum clearance and comprehensive redevelopment decades, and by no means only in Manchester (Marris, 1987). It should not be too much of a surprise that some of these projects are now having to be revisited, but we need to learn from the processes that have brought this about to try to discover patterns of public behaviour that reduce the likelihood of this being repeated in future. And we also need to reflect on the fact that much of this was the conventional wisdom of its time, and was staffed by well motivated and highly committed people; and so it manifestly is not enough to apply 20:20 hindsight to this and to apportion blame accordingly. We also need to recognise that, unless we are both thoughtful and careful in what we are doing and unless we do it in as inclusive a way as we can, we are in danger of failing to learn the lessons of the past which are not just about its artefacts but are also about its processes. All these things should tell us not merely that we need to do better but also that we need to do differently, and these understandings have been important components in the development of my thinking about seeing people not just as an amorphous mass but as clearly defined sets of customers.

Problems of terminology

When preparing the first draft of a paper for the 'skills and ethics' seminar at the University of Newcastle upon Tyne in October 1988 (which ultimately emerged as Kitchen, 1990, and as a chapter in Thomas and Healey, 1991), the term I used for both the intended and the unintended winners and losers as a result of the planning process was 'clients'. At that stage, I did not give too much thought to my choice of terminology; my focus was on trying to give expression to the approach as a whole. Subsequent experience has shown that this issue was worthy of more thought, not for semantic reasons but because there are subtle but important differences between the various terms that could have been chosen. Schon (1983, pp. 290–91), for example, argues that terms such as 'stakeholder' or 'constituent' are more appropriate in relation to planning activity than is the concept of it having clients. He was writing before the term 'customer' had any particular currency in these terms and so he doesn't debate this as a possible alternative, but it is interesting to note in this context that the idea of stakeholders has now come into more general use in relation to the Local Agenda 21 process (see Chapter 9). It has also been picked up and used extensively as a term by Hutton (1996) and in turn by the British Labour Party in expressing its philosophies for the late 20th and early 21st centuries, and as a consequence is in some danger of being turned into a term which means many different things to many different people. This matter of terminology has taken practical form in Manchester in recent times, with quite vigorous chief officer debates about how to describe the intended beneficiaries of the council's services in a situation where the council wanted to promote greater awareness and responsiveness in these terms and therefore felt (wholly reasonably) that an agreement on nomenclature would be a good starting point. The debate tended to polarise around three terms, each of which had its supporters:

- users
- clients
- customers.

There was probably least support for the term 'users', I think perhaps because it raises awkward questions both about people who happen at a particular point in time to be non-users but who would fall within definitions by the council of those at whom its services are targeted, but also because it doesn't adequately address people who are on the receiving end of council services such as legal action when that is designed to protect the interests of third parties. The debate between the terms 'clients' and 'customers' was much closer, however, with the majority view being that the term 'customer' probably had a less passive feel to it than the term 'client' and carried with it a much wider understanding with the general public of their entitlement to service; people understood being a 'customer' (because they had far more experience of it) much better than they did being a 'client'. It also encompassed a wider range of local government actions. For example, when taking legal action against someone to protect third-party interests, it is probably easier to think of the person against whom action is being taken as a customer with defined rights than it is to see that person as a client. The strongest argument against the term 'customer' is probably the implication it carries of someone who is willing to pay for something, but this becomes less of a problem if it is acknowledged that all services cost money, and thus the question becomes merely one of who pays directly and who pays indirectly. The strongest argument in favour of the term 'customer' is that it is a term now being quite widely applied to quite a lot of public sector activities, which are being encouraged to think of themselves not as being producer driven with a divine right to exist because they are part of the public sector but as consumer driven because they provide services that are needed, wanted and valued. Local government services ought in principle to be like this, and local government planning services are in these terms the same as other local government services. Consequently, it is not unreasonable that this sort of terminology should be applied to planning activities, although I can certainly recall some people taking umbrage at being so described face to face. I have therefore continued with this terminology here, rather than reverting to the term 'clients'.

The major customer clusters

There are many different ways in which the customers or clients of the planning service could potentially be defined. One obvious way would be to look at the make-up of the city's population and to define as customers the broad groupings there to be found (for example, women, children, elderly people,

people from ethnic minority backgrounds); and as Chapter 4 shows this was indeed an approach that was adopted as part of the work on the Manchester Unitary Development Plan. Leaving aside the point that the customers of the planning service in Manchester are not just the residents of the city, such a classification would also ignore the roles that are being played by customers; and it is this focus on roles that has determined the approach adopted in this chapter. It needs to be remembered, however, that within the structure that has been adopted here a whole series of other issues sit that would arise as a result of looking at the definition of customers in other ways. Thus, for example, the concern of disabled people about access to buildings cuts across the clusterings in this chapter both because several of the clusters would have a role to play in dealing with this issue and because disabled people could (and probably would) be found in every cluster defined.

I would now define 10 broad customer clusters for the planning service. In doing this, it is important to make the point that organisations can fall into more than one of these categories at different points in time; for example, cluster 4 can become cluster 1 on a particular occasion, and cluster 10 could in practice consist of organisations from most of the other clusters. The definitions, therefore, focus on the role that is being played, rather than on the organisation itself. The 10 clusters are as follows:

1) Applicants for planning permission.
2) Local residents affected by planning applications in an area.
3) The wider general public in an area.
4) The business community.
5) Interest or pressure groups in the community.
6) Other agencies whose actions impinge on the development process.
7) Other departments of the local authority.
8) The elected members of the council.
9) The formal control mechanisms of central government.
10) Purchasers of planning services.

The tenth of these (purchasers of planning services) was not defined when I first worked on this concept in written form in the late 1980s, simply because the service was not seen in those days as engaging in trading activities. I have already shown in Chapter 1, however, that this has been a marked trend of the 1990s.

It should be noted that not all these customers necessarily identify themselves as such, although a large number do. I have always taken the view that planning actions tend to have intended winners and losers, and also unintended winners and losers, and that as far as possible planners ought to try to identify who is likely to be affected by proposed actions and ought to try to ensure that those people are given the opportunity to get involved in the process and to express a view about it, even if they show no inclination to come forward of their own volition (see Reade, 1987). None the less, there are limits to this; for example, it is clearly circular to argue that unintended

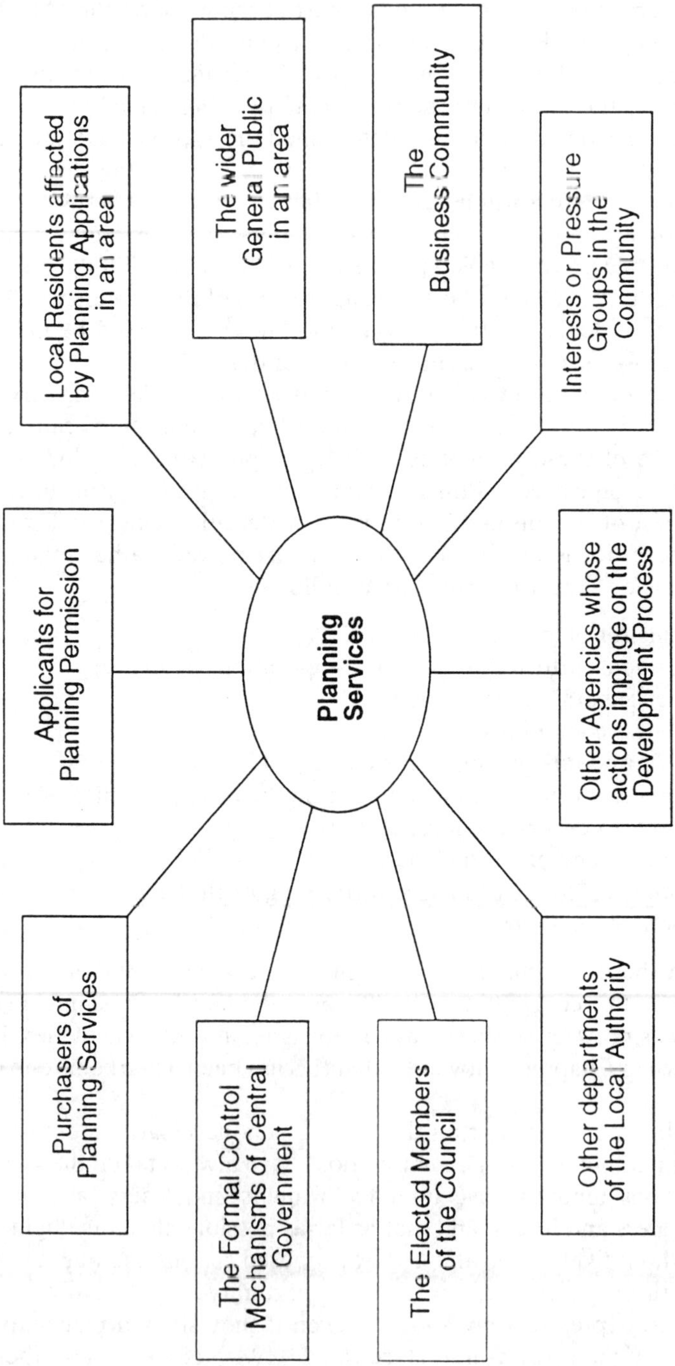

Figure 2.1 The main customers of the planning service

winners and losers should be identified and given the opportunity to get involved if the inability to identify them in the first place is what has put them into the 'unintended' category. The political process also has a very important role to play in all this in sanctioning definitions of who are to be seen as customers and in endorsing approaches to their involvement including the provision of funding to facilitate this. It is also worth noting that a lot of planner–customer relations evolve through the continuous process of interaction, and thus need to be seen not in the static sense of an individual case but in the dynamic sense of the ongoing process. All these points give rise to some difficult questions about what planners actually ought to do in practice to translate these precepts into reality, to which I will be returning at the end of the chapter.

Figure 2.1 shows in very simple form the relationships between the planning service and these clusters of customers. I have written elsewhere at some length about the characteristics of each of these groups (Kitchen, 1990). Some useful recent literature on some of the key customers which postdates that paper is provided by

- Bramley, Bartlett and Lambert (1995), on the private housebuilding process and the planning system.
- Adams (1994) and Ratcliffe and Stubbs (1996), on the property development process and the planning system.

This none the less remains an area which is still relatively underdeveloped in planning and related literature.

It will be noted that of these 10 customer clusters, two (other departments of the local authority and the elected members of the council) could be regarded as being internal to the council and the other eight could be regarded as being external. There is clearly some truth in this in the sense that planners and these two 'internal' clusters of customers should all be seen at any rate at some levels as part of the same council team. At the same time, this is a distinction that can easily be pushed too far, because of course each of these two could be analysed in terms of its own customer clusters. Such an analysis would no doubt show (as I have argued that it does for planners) that the majority of these customers too are 'external' on this basis, and it is this balance of customer orientations that is a prime determinant of the behaviour of both the corporate system of the council at officer level and of elected members of the council. I have gone into this in some detail in Chapter 3 in terms of how elected members impact on the planning process, because of the special nature of this relationship in a democratic society and because this has rarely been written about from first hand. But one of the necessary decisions about selectivity that has been taken in writing this book is that I have chosen not to include a separate chapter on planning and corporate processes. This is not an unimportant issue, and it is certainly not to argue that the planning process is able to operate in a free-standing manner untouched by and in turn not seeking to influence the work of the rest of the council; indeed, Part III of this book illustrates some of these relationships. But the level of detail necessary to do justice to these matters would deflect readers from the primary purpose of this book. These matters have been usefully written about elsewhere, albeit not in a specific Manchester context, in terms of how policy and financial issues are handled (Blackman, 1995; Fenwick, 1995; Stewart and Stoker, 1995), how they are affected by the changing nature of the relationships between central and local government (Duncan and Goodwin, 1988), how co-ordination and implementation issues get handled (Rein, 1983, pp. 59–77, 113–37, Greed, 1996), the particular position of the Chief Executive and the role of the Chief Officers' Management Team (Gyford, Leach and Game, 1989, pp. 110–18), and at a more theoretical level how the component elements of organisations relate both to each other and to the outside world through processes that are partly co-operative and partly competitive, described as 'partisan mutual adjustment' (Lindblom, 1965).

Relating to customers

The essence of the approach set down in this chapter is that the planning job is seen primarily to be about meeting as many of the needs of a wide range of customers as is possible, whilst acknowledging that the council both as the planner's employer and by virtue of its statutory powers does have certain rights of decision which have to be regarded as over-riding. This is obviously a very specific definition of the planning job based upon my experiences in Manchester, rather than an attempt at a universally applicable definition, and in particular it acknowledges the specific role of the council as local planning authority which is a feature of the British planning system.

There are obviously some tensions and difficulties in all this, the main ones perhaps being the following eight:

- An obvious argument for the customer-driven approach advocated here is the sheer mathematics of the situation. Some 440,000 Mancunians to 110 Planning Department staff creates odds of 4000:1, and most gamblers would not take odds like that because they would indicate that the chances of success were very remote. Clearly, therefore, it makes sense to try to reduce these odds somewhat by working in partnership with these customers, and perhaps in particular by trying to make use of their knowledge and understanding of very many of the city's component elements. The problem, of course, is that customers in these terms are far from being uniform, or consistent, or indeed at times even reliable. Thus, the desire to work with customers has to be tempered by a realistic assessment of their attitudes and motivations. It cannot simply be assumed, as is sometimes argued, that 'the customer is always right'.
- How can a balance be struck between serving the interests of the council and meeting as far as possible the needs of other customers? There clearly can be

some contradictions in this relationship, and where there are my experience is that they are best resolved by the planner being as straightforward and as honest as possible about the situation and his or her role in it and by seeking to respect and understand the positions and rights of other customers. This must include acknowledging openly the primacy of the council's position where it is likely to take an over-riding view, given that elected members in the democratic system of British local government are there amongst other things to make decisions. At the same time, it is possible to put too much emphasis on this as a source of difficulty. In my experience, elected members usually wanted to see matters resolved by agreement where possible rather than by imperative, and they also usually wanted participants in the process to feel that they had been fairly and reasonably treated.

- The need to ensure that there are clear links between on the one hand the strategic planning work that councils do in seeking to improve their areas and the quality of life of their citizens, and on the other hand the customer feedback that is obtained through the constant operation of the planning service. To put this more simply, policies cannot be ignored simply because a local head-count suggests that this would be popular in a particular instance, but equally the results of that head-count cannot just be ignored. There can be a problem of congruence in this simply because the strategic perspective may well be a broad and generalised one whereas customer feedback may be very specific, but the planning process has to be about producing the best connections that it can between these two. It must also be about ensuring that the council in undertaking strategic planning work is aware of what customer feedback is being received and how it may be related to strategic considerations, and this was an important factor in determining how to tackle the preparation of the Manchester Unitary Development Plan (see Chapter 4). Customer feedback must be one of the sources of continuous sustenance for the planning service.
- Are all customers equally entitled to receive a service, and in practice do they all do so? Unless the political process dictates otherwise (and, for example, it would be difficult to imagine Manchester City Council sanctioning the development of co-operative working relationships with a fascist group openly pursuing its objectives), the answer to the first part of this question must in principle be 'yes', but it is clear from practical experience that the answer to the second part is actually 'no'. There are many reasons for this that are social, economic or political in origin, but they mean that planners need to secure political sanction for and then actively pursue a strategy of being proactive in developing customer links. This perhaps relates particularly to the need to develop linkages with the most deprived communities in the city, rather than a planning stance of sitting back and waiting for the uneven and unequal process of customers presenting themselves to determine who gets services.
- Can all this be resourced? There is no doubt that this has become an increasing problem in recent years. A planning service with an establishment

of 179 upon the disappearance of the Greater Manchester Council in 1986 (although actual pairs of hands never quite got to this level) clearly does not have as many staff resources at its disposal as a planning service with 110 staff in post, as was the position for much of 1994/95, unless the functions to be performed have themselves been correspondingly reduced, which was not the case. Seeking to relate proactively to customers is itself a resource-consuming process, and it was very much the policy stance of the incoming Labour left administration in 1984 that we should do more of this. All I can really say in response to this is that at any point in time managers and staff have to try to do the best that they can in these terms with the resources at their command, and if for good reasons things cannot be done or services cannot be delivered it is best to say so openly.

- Is the form of organisation of the service sufficiently robust as to be able to sustain this approach over time? In the early 1980s, Manchester Planning Department had changed its organisational structure from one which emphasised functional groupings (so that, for example, there was a Local Plans Division, a Development Control Division and one that carried out environmental functions of various kinds) to one that sat multipurpose area teams alongside a more limited number of specialist groups, so that development control, local forward planning and local environmental improvements were all done or at least instigated by the same patch-based team. I was a passionate advocate of this approach (although not City Planning Officer at the time) because I believed that it enabled the service to relate better to its customers. In effect, it put the customer first. Despite declining resources, and in the full knowledge that there are some arguments that (for example) development control can be more efficiently carried out by dedicated teams, I always refused to countenance making structural changes away from this patch-based approach, because I saw such a move as sacrificing the interests of customers to the interests of a particular view of administrative efficiency. In saying this, I do acknowledge that within individual area teams the disposition of staff resources tended to work in slightly different ways according to the skills and abilities (and I suspect also sometimes the personal preferences) of individual members of staff, and I never regarded this as a problem provided it reflected a sensible approach to resource utilisation within teams at particular points in time. But the great strength of the patch-based team approach is the development of both an understanding of local circumstances, interests and concerns and the development of continuous relationships with customers, and as far as possible we did try to defend this within the financial allocations that were available to us. This does require, however, an explicit view about priorities, and such a view will not always be popular amongst staff who believe that as a consequence their interests or their work are losing out. This is what the language of priorities means, however; if everything is a priority, then nothing is a priority.
- Can planners actually handle the compartmentalisation of customer relationships that this approach implies, whilst at the same time both retaining

the respect of customers and maintaining the ability to give sound advice to the City Council? My experience is that, whilst this obviously varies from individual to individual, the broad answer to this question is not only 'yes' but also that quite a lot of patch-based planners actually get a great deal of job satisfaction out of this part of the process.

- Can planners handle the need to relate to customers in ways that respect customers' own aspirations whilst at the same time acknowledging that they are themselves likely as individuals to have (at times very strongly held) views about what actually ought to be done in particular situations? Planners are not neutral or value free in what they are doing. To operate effectively, they have to accept that there is a place for their own views but also a place for continuous learning based upon customer feedback and upon ongoing relationships which can lead to the modification of those views. In any event, of course, the council as local planning authority when it has a decision to make needs not only to know what the advice of the planning officer is but also to know what are the views and aspirations of interested parties, which the council may well decide should have primacy in a particular case. Not all planners are equally good at handling these tensions, but the ability to operate on this basis sometimes distinguished staff who were likely to rise to the more senior positions in planning organisations from those who might not be able to. If planners are to handle these matters well they require an acute awareness not merely of what they are doing in a particular set of circumstances but also of what they are perceived by others as doing. Many people will accept that the outcome in a particular set of circumstances may not be the one they would have wanted provided they feel that they have received fair and reasonable treatment in the process, with their needs, aspirations and positions respected, understood and properly reported by planning staff. Continuing relationships can survive difficulties over individual cases in these circumstances, but they struggle to flourish in situations where people feel that they have been manoeuvred or given partial treatment, and this can be at least as much about perceptions of planner behaviour as it is about the behaviour itself.

Securing customer feedback

The single most important source of feedback is the continuous process of interaction between planning staff and the customers of the service through its day-to-day operations. If these sorts of relationships are not working well, in Manchester's case on a patch basis via the area teams and with interest groups through the specialist teams, other forms of feedback are going to struggle to replace them. There are none the less at least four other sources that are valuable in their own right:

- Purpose-constructed consultation exercises.
- Through the political process.

- Through the local media.
- By asking customers direct.

Later chapters deal with the political process as a source of feedback (see Chapter 10). Purpose-constructed consultation exercises are in a sense self-defining (see, for example, the discussion on public consultation on the Manchester Unitary Development Plan in Chapter 4), although that does not mean to say that they are always successful. For example, throughout much of the second half of the 1980s, Manchester departments held a series of working party meetings with representatives of relevant customer groups to discuss the development of equal opportunities policies and practices. These were really types of focus groups with particular customer services, but as far as the planning service was concerned we frequently found it very difficult to get much useful customer feedback by this method, and it was eventually abandoned. One of the most frequent experiences of undertaking these sorts of activities, however, is that irrespective of the purposes for which they are constructed feedback is often obtained on all sorts of other things as well. This should not be ignored simply because it does not fit the exercise, but should be directed at the staff (sometimes in other departments) to whom it does relate. What is perhaps more unusual for a planning service is asking customers directly for feedback about the services they receive and don't receive, although the judgement that this is still quite an unusual phenomenon of itself speaks volumes about how far there is still to go in building customer-driven perceptions into British planning practice.

During the early 1990s Manchester City Council carried out quite a large-scale exercise in securing public comments about service delivery through a MORI survey of residents, which *inter alia* included some comments about the planning service. During the early part of 1993 the Planning Department went beyond this and carried out a customer satisfaction survey in respect of applicants and agents and the development control process, by sending out around 600 questionnaires to all the people who received decision notices over the period in question. All sorts of things can be said about the statistical dimensions of this approach, such as the response rate achieved (which was about 19 per cent with no follow-up of non-responses); but none of these are the point being made here. What this survey did was to ask this segment of our customers what they felt about dealing with us and how that compared with parallel dealings that they may have had with other local planning authorities. What this actually told us was that it was worth trying to do something like this because some sort of response was capable of being secured; that our broad philosophies as a department were on the whole being put into practice as measured by their experiences of us; and that our customers were prepared to make comparative comments.

The most encouraging comment in terms of the translation into practice of our philosophies was that just over three-quarters of respondents agreed or strongly agreed that the department 'generally tried to help people like me'.

As far as their overall view of our service was concerned, on a comparative basis with other local planning authorities with which they had dealt, 50 per cent said that we were much better or slightly better, 38 per cent said we were average, and 12 per cent said we were worse or much worse. The results weren't very different when compared not with other authorities but with respondents' own expectations once non-responses to this particular question had been eliminated; 45 per cent thought that we were much better or slightly better, 47 per cent thought we were average, and 8 per cent thought we were worse or much worse. Whilst as reported above the development control service was generally regarded as being helpful, it was also generally regarded as being too slow; for every seven respondents who supported this position, only four disagreed with it. There tends to be an inverse relationship between trying to be helpful and trying to be quick which is at the heart of managing the development control process (see Chapter 5), but it is clear from the results that our customers wanted both these things.

In a sense, these results confirmed our own preconceptions from the other sources of feedback that we were receiving, but it was important none the less that this was done and we were of course open to the possibility of receiving a rather ruder shock than this. The point to be made is that it is possible to get direct customer feedback of this kind and that it is worth doing; but departments have to be prepared to resource the processes of carrying out such a survey at all stages, and then to listen to and to think about the results that are generated.

Much of the specific experience from Manchester on which this chapter reflects is also visible more generally in the recent study that has been done for the Department of the Environment by McCarthy and Harrison (1995) of attitudes to the town and county planning system and service in Britain. The main conclusions of this are as follows:

- The general public has a good level of understanding of and a broad measure of sympathy for the development control process, but understands other aspects of the planning system (of which it sees property developers as the main beneficiaries) much less well.
- Three-quarters of the sample of applicants for planning permission recognised the planning system to be valuable, and less than 5 per cent saw it as having no value. Generally, planning application costs, speed of decision-making and the fairness of the outcome were all regarded as being acceptable, although there were some individual sectors which expressed concerns about one or more of these matters.
- Business generally has a much more sophisticated understanding of the planning system than either the general or the applicant publics, and a much wider range of views about it as a consequence. Businesses tended to be more critical of the system as a whole than they were of its handling of their own applications, with a widespread view being that it is arbitrary, expensive, slow and unresponsive. There appears to be an element of

received wisdom in this, however, given that on the whole they did not report their own experiences in such severe terms.

- Developers and landowners reported a very wide range of performance by local planning authorities, from those providing very highly regarded services to those seen as obstructive, dogmatic and unfair. Great stress, however, was laid on the value of early access to officers for discussion and negotiation, and overall (although there was a range of opinion about this) these processes were seen as improving the quality of development.
- Non-governmental organisations also report on this wide range of performance by local planning authorities, but in addition comment on the difficulties that local planning authorities sometimes experience in balancing the range of interests on issues and on the problems that can arise in terms of the cumulative impact of decisions from a system that prides itself on considering each case on its merits.

Conclusions

I have tried to set out in this chapter a view of planning as a customer-driven activity which is significantly different from the 'public interest' perspective that was the dominant view when I and my generation of planners were students. Whilst it has some theoretical components to it, it is rooted in practical experience over many years. Acknowledging that the planning service has multiple and overlapping customers, and that it is dealing with an infinitely complex and dynamic set of relationships, are both important parts of this. Perhaps in particular, the realisation that the planning process has stewardship responsibilities in terms of the development of our cities, but that they are cities which belong to their people rather than to their planners, is central to this perception of the planning job. With the amount of knowledge, experience and understanding of the way the city works that collectively exists amongst customers beating that available to planning staff in a ratio of many thousands to one, the ability to tap into this resource both adds to the information available to the planning service and also generates more cautious views of the fallibility of that service than has sometimes been the case in the past. In this latter context, being constantly reminded that the planning process in previous generations resulted in 'planning failures' which affect to this day the way people often relate to planners does have the effect of concentrating the mind. At the same time, the service can and does perform useful and necessary functions, is able to bring resources of skill, knowledge and understanding to bear which add value to what would otherwise be available, and can operate in doing these things in ways which are valued by customers. This can coexist alongside the process of advising the City Council in its capacity as local planning authority. There are no easy answers to the questions of how to handle the tensions and contradictions that can be found in this situation, but acknowledging that they exist, being clear about what one is trying to do and why, and being as open and as straightforward with customers as it is possible to be on the basis that such actions will be part of a continuing relationship all represent good starting points.

Indeed, whilst generalising from this experience about how to handle all these customer relationships effectively is an exercise fraught with dangers, I would argue that there are at least seven characteristics of planner behaviour which are essential to this:

- We have to be seen as competent professionals by the people we relate to. This means being able to bring both substantive and procedural knowledge to the table which adds to what is already available to our customers, and to communicate it effectively to our customers in ways that they can readily understand.
- We have at all times to be seen to be behaving with integrity. If we say something to someone, we have to mean it; and we will be undermined if we then go off and say something different to someone else or do something different, because eventually these various customers will become aware of this.
- We have to give everyone a fair crack of the whip, and we have to carry conviction that this is what we are doing. If we show people that we will not take their views seriously, and do not ensure that these are properly reported and genuinely taken into account when preparing our analyses, we will not pass this test.
- We have to be willing and able to listen. We don't necessarily know best, and we have to be open to the possibility that others may have persuasive or compelling points of view. A willingness to listen is a primary requirement of this.
- We have to be flexible in our thinking and in our behaviour. If we don't have flexibility in these terms, we will not be able to see how other people's inputs can be turned to constructive advantage, and we will be limited by our own frames of reference.
- We have constantly to be looking for opportunities to help to meet customer needs or aspirations where this is possible. People often know what they want in a very general sense only, and we have to be willing and able to help people to translate this into specific ways forward.
- We have always to be looking to build relationships with customers on the basis of a degree of continuity. Many of the customers of the planning service will be so on a recurring basis, and thus relationship-building is an important part of what we have to do even though people may initially present themselves on the back of a specific case.

None of this is to deny that in the final analysis the responsibility of a planner employed in local government in Britain is to the council which employs that individual. This is not merely an argument about loyalty to the paymaster. It is also a reminder that in British planning law it is the council that is the local planning authority, not the individual planner; and the job of the planner in that situation is to help the council to discharge those onerous responsibilities as best it can. At the same time, it should be remembered that councils exist to provide a range of services to their citizens, and the planning service is one of those services. Thus, councils as local planning authorities have a direct interest in achieving the highest levels of customer satisfaction that they can in providing a planning service, within a framework for service delivery constituted by their policies and practices. Much of the rest of this book is about the interactions between this framework and the customers of the planning service.

3

Working with elected members

Introduction

The traditional functions of chief officers are to advise, to implement and to manage. These come under particular strain with the intense scrutiny that is undertaken by the elected members of the council. The purpose of this chapter, therefore, is to look in more detail at the roles that elected members play which impact upon the planning process. The perspective adopted is that of a planner looking outwards at what elected members actually do. In this sense, it is complementary to the approach of Blowers (1980, pp. 8–38), who looks at the relationships between planners and politicians from the perspective not merely of an academic but of an active politician who for a time had been a committee chair. His focus is also essentially that of someone participating in a two-tier process from a base in the upper tier, whereas the focus here is essentially that of unitary local government (the situation in Manchester since 1986).

It is clear from the literature that there are several ways of looking at the functions that councillors perform in British local government. Gyford (1984, pp. 13–24) uses a 13-point typology, having broadly split local councillors into 'tribunes' and 'statesmen'. This is summarised in Box 3.1.

Newton (1976, pp. 114–44), based upon his study of the politics of Birmingham, sees elected members as having six role orientations: the representational role; the party political role; a role in relation to geographical areas, which may well be wider than their own wards and encompass the whole city; orientations towards policies or towards cases; topic specialisms, in some cases in relation to particular parts of the council's work; and roles in relation to pressure groups. He also sees them as falling into five role types, which he describes as parochials, people's agents, policy advocates, policy brokers and policy spokesmen. Stewart (1983, pp. 74–78) classifies members according to their local political seniority into the leadership, service chairs and backbenchers. Wilson and Game (1994, pp. 204–27) see the roles of councillors in terms of what they actually do, with the main roles being those of representative, of policy-maker, of manager and of monitor and progress chaser. Elcock

Box 3.1 Gyford's 13-point typology of the determining characteristics of councillors in British local government

- Their status ranges from junior to senior.
- Their ward types range from marginal to safe.
- Their political styles range from delegate through politico to trustee.
- Their focus can be on their ward, on the wider community or mixed.
- Their relations with their constituents can be those of welfare officer, of communicator, or of mentor.
- Their relations with interest groups range from being facilitator and spokesperson to being resistor or referee.
- Their issue orientation can be general or specific.
- Their form of involvement can range from casework through to policy.
- Their relations with officers can range from acting as a watchdog to being a colleague.
- Their information sources can be internal or external.
- Their behaviour in relation to information flows can be passive (wait) or active (search).
- Their political views can range from ideological to administrative.
- Their political behaviour can range from maverick to loyalist.

(1994, pp. 61–108) looks particularly at the relationships between councillors and officers, and then extends some of that discussion into a particular examination of planning from the perspective of managing uncertainty (*ibid*., pp. 233–64).

The approach adopted here is not to work explicitly from any of these particular classifications, since they are not mutually exclusive, but rather to illustrate and extend them by looking at the different roles that elected members play which impact in practice upon the operation of the planning process. To this end, the nine roles performed by elected members in relation to the planning process are listed below, and then each is looked at in more detail. In particular, it should be clear that this chapter is not based upon the 'dictatorship of the official' view of local government, which was at one time common in the academic literature but is now more doubted (Gyford, Leach and Game, 1989, pp. 95–160), but takes the view that elected members play major and worthwhile roles.

The nine roles of elected members

The nine roles actually played by elected members that I can differentiate from experience of how they impact on the planning process are summarised in Box 3.2.

It is possible for members to play one, some or (in the case of members of the political leadership) all these roles, at the same or at different points in time. It is also worth noting that some of these roles can be in conflict with each other, and where individual members perform conflicting roles they have either to resolve these conflicts for themselves or to behave like chameleons in different settings. It should be noted, however, that this is an analytical construct. Elected members do not actually make these distinctions for the most part, and experiences and contacts gained when performing one role overlap into others and form the framework for their future performance. None the less, it perhaps helps to understand how elected members in their various ways impact upon the planning process by trying to separate out the various roles that are performed in this way.

Box 3.2 The nine roles of councillors as they impact upon the planning process, drawn from Manchester experiences

- as ward representative;
- as members of a political party;
- as members of the leadership of a political party, and in particular as members of the leadership of the majority group on the council;
- as chair and deputy chair of the Planning Committee;
- as members of the Planning Committee;
- as members of another committee whose work impinges on the work of the Planning Committee;
- as members of the council, particularly in the wider sense of its role as representative of the city;
- as protagonists of individual policies or views, often working in partnership on these matters with relevant interest groups; and
- as individuals in their own right, with their own interests and particular behaviour patterns, and with their own personal ambitions.

Members as ward representatives

All members of the council represent wards of the city, and this is their base. Their level of interest not merely in what is going on in their wards but also in what active people and groups in the ward think about it is thus usually relatively high. It is also often very conservative. The reason for this is that people often express their dislikes or their fears about change proposals, like for example planning applications, but do not tend to express positive or neutral views about such proposals when they have them anything like as frequently. Since elected members generally like to be seen to be supporting the sorts of views that are being expressed in their wards, on planning matters and particularly on development control this often involves speaking against something or writing a letter of opposition about it even when they know that in terms of the policies that they have approved it is likely to be acceptable. When asked about this, members will typically say that they are aware of these circumstances but have to be seen to be helping local interests at least by ensuring that their views are known as part of the decision-making process. Occasionally, these sorts of views might prevail in an individual instance, perhaps because a ward member has obtained sympathy or support from Planning Committee members, and then the question of whether this might be the beginning of a policy shift or is merely an individual decision which does not carry such wider implications needs to be thought about (and usually discussed quietly with the committee chair); but generally speaking this is known, accepted and understood ward behaviour by members who wish to be seen to be offering local support, which most members acknowledge that they themselves are likely to need to engage in from time to time without expecting that the Council will take aberrational decisions as a result.

This isn't to say that all elected member involvement at ward level on development control matters is negative, and some members are enthusiastic supporters of the process of negotiating improvements to a submitted scheme, keeping in touch with staff regularly throughout this process for progress reports. In addition, ward members will often play a very useful role in introducing and working with local people or groups who are interested in promoting actions such as local environmental improvements. The function they perform as interfaces between such people or groups and the professional staff of the department can be very valuable in helping people into what must appear as the labyrinthine complexity of local government. What typifies much of this, however, is activity in the local community, which is very often in turn the trigger for local member activity.

This is not the only basis on which local members express ward views, however. Some members personally hold very strong views about their locality and have a deep local knowledge also, and thus will take steps to try to get things done irrespective of whether others in the community are seeking to do likewise. This is usually likely to be via their knowledge of how political and officer-level decision-making processes operate and how best to access them, rather than via public statements, although sometimes these coexist.

What ward members want more than anything else in all this tends to be information. They want to know in terms of the council's functions what is going on in their ward, and they don't like being taken by surprise for example when a constituent comes to them with a problem or issue at one of their surgeries. To this end, ward members frequently make a point of establishing and maintaining contact with the patch-based planners working in the area teams whose areas contain their wards, and the two-way flow of information to which this can give rise can be very helpful. Sometimes, particularly when all three of the councillors from a ward come from the same political party, one of them will specialise in planning issues and will become much better known to the staff as a consequence than the other two. Sometimes also a high level of intensity of planning activity in a ward can lead to a decision by elected members from it that they want to come on the Planning Committee, although this isn't the only basis on which members make such decisions. As an example, when Labour took Didsbury from the Conservatives at the 1994 local elections, the new Labour councillor for that ward came straight on the Planning Committee, thus maintaining the representation from Didsbury on the committee that had been provided for many years by the stalwart Conservative member that he replaced. The reason for this was that in Didsbury ward, with its large number of articulate and well organised people, planning was an important matter; indeed, in terms of the kinds of views being expressed by its member on the Planning Committee the change from Conservative to Labour did not make an enormous amount of difference, and its new councillor still tended to speak in support of local views irrespective of party labels. Thus, Didsbury councillors needed to take an active interest in planning issues, and securing a place on the committee was perhaps the most straightforward way of doing this.

Councillors as members of political parties

In Manchester, all the members of the council get elected through the party political process. There are examples of people who have fallen out subsequently with their parties, been cast into the political wilderness by those organisations or voted with their feet and then survived in effect as independent members until they are due for re-election; but even where they have then stood in their own right to defend their seat it has been against a candidate from their own party and the outcome has been a heavy defeat. Thus the party political process provides the effective frame of reference for council membership and activity. It should be said, however, that in Manchester there were not for the most part strong differences of opinion along party lines on planning matters, at any rate not between the Liberal Democrat group (which has formed the 'official opposition' in recent years) and the Labour group. Thus, whilst debate about an individual issue might polarise on party lines, it could not really be said that committees operated on the basis of strongly differentiated views on planning matters between the parties. There was often a consensus on a high proportion of committee agenda items, and a willingness to accept a range of points and views in discussions from across the political spectrum.

That having been said, the parties are very variable as to how much of a constraint this is in practice on what elected members actually do. Party policy is usually the starting point for this, and parties usually take very seriously their own firm policies, to the extent that they have them. But the point has already been made that elected members very often want to be seen to be supporting local representations even when this is against declared policy, and the rights of members in this regard tend to be safeguarded in effect by creating holes in the rules or practices by which the parties abide. This usually involved a distinction being made between the right to speak and what actually happens when votes are cast. For the Labour group in particular, voting against something which has been discussed in the group, a majority view decided, and in effect (and sometimes actually) a whip imposed is a serious matter; but seeking the right to speak and then becoming invisible when formal votes are taken in committee is not uncommon. It is also important to note that individual members can often persuade their group to go along with a particular view even though it may not be consistent with policy; and as has already been said, this can sometimes be the trigger for a call for a policy to be reviewed, although it is important not to be confused between a genuine policy question that may need rethinking and a political decision that is not intended to challenge the policy framework. This can in turn cause difficulties for planning officers because of the danger of individual case decisions of this kind subsequently being quoted as precedents by people anxious to see a policy shift; but there is usually enough in the particular circumstances of each case to avoid too many problems arising from 'political' decisions, and usually also enough awareness of the potential difficulties that adverse precedents can cause amongst committee members (and particularly the chair and deputy) for this point to be taken into account in their discussions. Members will sometimes support each other in these circumstances, because they know that there may come a time in future when they will in turn be looking to their colleagues for support. It should also be said in this context that policies don't always provide a clear-cut basis for decision-making. They cannot anticipate every circumstance, and a lot of cases raise several policy issues which don't in any event all pull in the same direction. A lot of planning casework is like this, and thus there is often room for members to persuade their colleagues that a particular decision is right in a particular case without necessarily placing a policy at risk; a decision, for example, will be regarded as a justifiable exception rather than a precedent. All of that having been said, however, it is important not to get this out of context; the vast majority of decisions go with recommendations which are based upon existing policies, and the sorts of changes being discussed here occur in a tiny minority of cases.

What does become particularly significant, however, is the view that a party group takes as a whole about an issue. The Labour group, for example, sometimes uses the mechanism of deferring something that is known to be particularly contentious to the council, in order to enable a debate about it to take place in the full Labour group which meets the night before the council. The majority decision of this group meeting produces a whipped view for the council, which can often then go through at council without any discussion at all. There is, of course, a sense in which the views of the full group are the collective views of each of its individual members, but whilst each of them has one vote in the group meeting it cannot be said that this means that the views of each member are therefore of equal weight. In practice, the views of the political leadership of the party group or the chair and deputy chair of the committee (or in the case of the minority parties, the committee spokesperson) that is raising the issue are likely to carry the day at the group, although there certainly are individual cases where members have successfully lobbied to defeat views emanating from these sources. Because the wishes of either the political leadership of party groups or of committee chairs and deputies usually prevail in these circumstances, these are the roles that are dealt with next.

This section has been written from the perspective of someone who has worked for an authority where consistently the Labour party has had a very large majority. Opposition groups have been small, with the Conservative Party in recent years being reduced to the point of virtual non-existence and the Liberal Democrats growing relatively slowly and concentrated into particular parts of the city. As a consequence, much of the real opposition has come from within the Labour group; its factions and its rivalries have often raised more policy challenges than have the conventional processes of dispute between political parties. It should be recorded, however, that when I have tried out the categorisation used in this chapter at conferences containing elected members in the audience, I have sometimes been told that it does not deal adequately with situations where there are sizeable opposition groups or indeed where no party has overall control of the council. The reason for this, as I have said, is that these circumstances have not obtained in Manchester, and thus I cannot talk about them from first-hand experience. I am wholly prepared to accept, however, that my analysis of how councillors as members of political parties operate in relation to planning decisions would need to be extended in situations where bargaining between political groups plays a large part in decision-making because no single group has enough votes to guarantee that its point of view will prevail, or in situations where sizeable formal oppositions have a well worked-out and distinctive philosophy in relation to planning matters.

Political leaderships

It is not very easy to generalise about this particular role, because in practice during my time as a chief officer in Manchester the only changes in these terms were in the leaderships of the minority parties. There isn't much help from the available literature either, because local political leadership doesn't appear to have been analysed in a great deal of depth, and what has been written places a great deal of emphasis on the particular character and characteristics of the individual leader (see Stone in Judge, Stoker and Wolman, 1995, pp. 96–116). More generally, Judd and Parkinson (1990), through an international set of case studies including Glasgow, Liverpool and Sheffield from Britain, show that leadership can make a difference to the outcomes of urban regeneration initiatives. There are two important caveats to this, however. The first is that effective leadership does not always come from the local political sector, but can come from the local private or business sector or from cross-sectoral working. The importance of the local political leadership's actions, perceptions and abilities to the successes (or otherwise) of the stories told in the three British case studies does appear to show, however, that in Britain at any rate local political leadership is likely to be at the heart of urban regeneration initiatives. The second caveat is that the definition of success in the various case studies assembled by Judd and Parkinson appears to be quite a narrow one, and they conclude (*ibid.*, p. 307) that leadership may make a considerable difference to whether (and how) a city regenerates its economy but makes much less difference to the issue of equity. The question of how to use economic regeneration as a means of improving the social and economic circumstances of the most deprived sectors of the populations of cities is shown not to have been very effectively tackled even in the most 'successful' of the regeneration cases examined. This conclusion is also broadly shared by Wolman and Goldsmith (1992, pp. 184–222). This echoes the debate about these same problems introduced in Chapters 6 and 7 of this book.

Graham Stringer had been Leader of the Council since 1984, and although the composition of the Labour group officers supporting him in this role changed over that period it did not appear to make much real difference to the trends that were visible in the directions of his particular brand of leadership. In saying this, however, it shouldn't be assumed that the process of political leadership only relates to one person. The phrase tended to be used loosely by people to describe the small groups working at any one time closely with the Leader of the Council, including those people who were formal office-holders in the council's Labour group. He was the fixed element, however.

There were three trends in particular that to my mind marked Graham Stringer's period of leadership, which on the whole would have been regarded by the outside world as a very successful period for Manchester by virtue of the scale of the city's urban regeneration achievements during that period which are catalogued in this book. These trends became particularly dominant in the later years of his leadership:

- The emergence of a specific agenda that was radically different from the cluster of causes and commitments that accompanied the change of power within the Labour group in 1984 from old centre-right to the new left. This emerging agenda focused on economic development mainly through large projects (the Concert Hall, the Olympic Games bid and the Commonwealth Games bid were all examples of this), on working in partnership with the private sector and making sure that partners' concerns and interests were met, and on job generation. It was an agenda that was pursued more obviously after the 1987 general election, when it had become clear that Thatcherite Conservatism would still be around for some time with its own doctrines about local government. Thus it was at least in part a pragmatic response to the long-term existence of a right-wing Conservative government, designed to try to get the best for the city that was seen to be attainable in this situation. It was also an agenda that became increasingly resistant to new thinking that might challenge some of its basic precepts, as the story of how the emergence of air-quality issues was tackled (see Chapter 9) illustrates. Depending on who was describing the scale of change over the decade or so that this encompassed, it was either a sharp shift to the right or a parallel journey to that being made nationally by the Labour Party. Whichever of these views is accepted, however, it was certainly a considerable feat in adaptation.
- A process of centralisation was clearly visible, despite the firm commitments in the early years to decentralisation and to enabling and empowering more

people. Committee arrangements that in practice gave more power to the Policy and Resources Committee (which is chaired by the Leader of the Council) and through it to a raft of subcommittees exemplify this, as for example the story of committee arrangements to oversee regeneration activities in Hulme (later largely mirrored in north Manchester) told in Chapter 7 illustrates. This visible process of centralisation was accompanied by an invisible process, whereby more was done through the Leader working in close conjunction with some staff in the Chief Executive's Department. From time to time this was the subject of muttering in the Labour group by members who felt effectively excluded from many decisions, but it didn't often become much more than muttering. This was an agenda that was about control, as the Leader's grip on the reins of political power tightened.

- An associated process of 'deification' was also visible. The young new Leader in 1984 who appeared to be happy to preside over a ferment of ideas from a coalition of people revelling in the opportunity to take initiatives had become someone by 1995 whose approval was apparently essential for any new initiative. This was also a self-fulfilling process, in the sense that staff particularly in the Chief Executive's Department often treated the Leader with excessive respect. 'What does the Leader think about this?' became a more important question than 'What should be done about this?' It almost certainly meant also that the claim that something represented the view or the wish of the Leader became a powerful tool in arguments. I suspect that this was sometimes used as a weapon even when the Leader had neither expressed such a view nor even been asked about the matter in question, such was the significance attached to something purporting to be the view of the Leader.

Much of the credit for Manchester's regeneration efforts in the late 1980s and early 1990s properly belongs to Graham Stringer. I think also that he rightly deserves credit for building effective partnerships with some elements in the private sector where previously relationships all too often had been characterised by hostility. My experience, however, was that the elements described above became more dominant as time went on, particularly in the last few years. Furthermore, this was an analysis that was quite widely shared within the Labour group, if private conversations with many of its members over the years are to be believed. This never really became the basis for a challenge, however, partly because this was such a daunting proposition in the later years but partly also because many Labour members were proud of what the City was achieving during this period and happy to be part of an administration with this growing list of achievements to its credit. The continuation of the sorts of processes described above to the point at which a leader is seen as being both all-powerful and permanent can, of course, breed some undesirable consequences; see for example Lawson (1992, pp. 282 and 305) who uses the medium of satire to ridicule what can happen with long-term and all-powerful leadership.

All this tended to mean that life for a particular service could be made very difficult both by the direct actions of the Leader of the Council and by their follow-up via instructions to the Chief Executive's Department if that service became the subject of the Leader's displeasure. Given the nature of his personal agenda which has been described above, the planning service was always likely in any event to be drawn to the Leader's attention because it dealt directly with things which impinged on that agenda; planning was not one of those departments that had very few dealings with the Leader because its subject-matter wasn't of particular interest to him. It has to be acknowledged as well that there were occasions when he had come across the work of the Planning Department or of individual members of staff where things had clearly not gone as well as they could or should have done, and he tended both to be an astringent critic in these situations and to build sizeable generalisations on the back of these experiences. Indeed, his view of the planning service seemed often to be a negative one because he saw it as obstructing what he was trying to achieve, although I was never very clear whether this was a judgement based upon unambiguous evidence or a preconception reinforced every time he came across something that appeared to fit into it.

This was compounded in the case of the Planning Department by the growing policy tensions between Graham Stringer and Arnold Spencer in his capacity as Chair of the Planning Committee, some of which are chronicled in Chapter 9. My naive hope in late 1992 that I might be able to build some bridges here in the immediate aftermath of my return from being Acting Chief Executive was quickly abandoned in the face of unmistakable evidence that this was territory where officers ought to fear to tread. These difficulties were in turn given a public face by the *Manchester Evening News*' tendency to cover extensively the words and thoughts of the Leader, a process that he undoubtedly encouraged (although almost certainly not in all cases) by the access he granted to certain journalists and by his ready availability for quotes, photographs and all the usual paraphernalia of a politician anxious to create a positive press image. The nadir of these processes was the blazing headline 'Stringer Blasts City's Planners' on the front page of the first edition of the *Manchester Evening News* of 25 January 1994, and then (just to make sure that we hadn't missed the point) 'Stringer Lashes the Yes Minister Planners' on the front page of the second edition of that same day. The point about this type of use of 'the public pillory' (to use the *Manchester Evening News* apposite phrase from an editorial of 26 April 1995) that seemed particularly unfair was its one-sided nature. The Leader of the Council did things like this, but there was no way in which I as an employee of the council irrespective of how strongly I felt about the unfairness of this could reply in kind. Quite apart from the reactions of the department's staff to this sort of treatment, it undoubtedly also created a problem for the service in its relationships with its customers, many of whom were also *Manchester Evening News* readers. The fact that these sorts of things were happening also affected internal relationships within the council, and showed how quickly things could change; after all, 16 months before the appearance of the January 1994 articles I had been the council's Acting Chief Executive.

It is wholly possible that these dismal, and relatively recent, experiences of being on the receiving end of the council Leader's displeasure have led me to paint the process of relating to the political leadership in darker colours than the overall record would justify. There is much that was positive for the City of Manchester that occurred during his period of leadership, and this book deals with these achievements for which Graham Stringer deserves much credit. And I must also record in this context that until the events of January 1994, I had thought that my personal relationship with the Leader was a good one. On the other hand, I know from first-hand experiences and confidences that the process of consolidating ever more effective power into one pair of hands was a source of concern to many officers and members inside the council, including members of the Labour group who collectively had allowed this to happen. Manchester City Council operated a very sensible rule whereby committee chairs could only occupy that position for four years in order to limit the scope for the creation of a prolonged personal fiefdom, but this did not apply to the Leader of the Council (who by virtue of that role was also Chair of the Policy and Resources Committee). When it was announced that he would be standing down as Leader after the May 1996 municipal elections in order to pursue his Parliamentary ambitions, he would have been Leader for 12 years or three times as long as other members were allowed to be chairs. It was no surprise to find him advocating in Labour Party circles the idea of a big-city mayor elected at large and with extensive powers to act (including those of hiring and firing staff), and with the council reduced to a much more limited role, along the lines of the model that operates in some US cities. Arguably, this was a natural extension of the powerful position he had been building for himself over the years within the existing system anyway.

Much of the discussion in this section echoes what Benveniste (1989, pp. 189–93) has to say about the apex of political power. He uses the term 'the Prince' (drawn from Machiavelli, without necessarily implying any of the other overtones that would be commonly associated with this source) to describe this location, accepting that in some circumstances this may be an individual and in others a collection of people, and he shows both how the Prince operates and how difficult the relationship can be between the Prince and his lieutenants and the planners. His summary of the process chimes well with what has been written above (*ibid.*, pp. 192–93):

> Meanwhile, his planners cannot take him [i.e. the Prince] for granted. His support will have to be sought, and the advantages that will accrue to him will have to be made explicit. The planners will have to entice him to acquire his trust. They will use his name and act under his authority even when they have not revealed what they are going to do for him. He will close his eyes and pretend he does not see what is going on. He will be ready to jettison the exercise whenever the heat becomes too intense. The planners will provide a shield for his mistakes. He will claim he did not know; he will take refuge in deniability and survive while they take the blame.

Chair and Deputy Chair of the Planning Committee

A city planning officer has to work very closely with the chair and deputy chair of his or her planning committee. This axis determines to a substantial extent what the executive and the political arms of the local government planning service do, and in this sense the relationship is symbiotic; see Wolman and Goldsmith (1992, pp. 158–63), for a general summary of the academic literature on this matter. But it must also be robust. Planning officers have to be able to give advice which does not work backwards from what they might believe the political process would like to hear, but which is the best appraisal they can give of the most appropriate way forward in a particular set of circumstances. Equally, chairs and deputy chairs need to be receptive to advice of this kind even if its content may be unwelcome, and they then need to balance this against other considerations (and particularly political ones) that they judge are also relevant to these circumstances and to how they intend to respond to them. A relationship of this kind needs to be worked at. It doesn't just arrive overnight merely through the assumption of roles, but builds over a period of time as mutual respect grows through demonstrating that the executive and the political strands by working together can each add value to the other.

I was very fortunate during my period as City Planning Officer and then later as Director of Planning and Environmental Health in having a continuity of talented people in these posts. Andrew Fender and Arnold Spencer were my two committee chairs during that time. Andrew had previously been Deputy Chair of the Committee, and when he finished his period as Chair he reverted to Deputy Chair and also became Chair of the Development Control Subcommittee. Arnold had previously been Chair (he was the first Chair when the left took over in 1984), and returned in that capacity when the Highways Committee (of which at that time he was Chair) and the Planning Committee were combined for a time in the early 1990s. He remained in that position when the Planning Committee became the Environmental Planning Committee in 1993 until he lost his position at the Labour Party and group elections immediately following the 1995 municipal election, after a series of very public disagreements with the Leader of the Council. These changes of committee names and responsibilities, incidentally, had on one occasion the effect of keeping Arnold Spencer in office when under the four-year rule (a council standing order which means that committee chairs, except the Chair of the Policy and Resources Committee who is also the Leader of the Council, stand down from that post after four continuous years in office) he would no longer have done so.

Both Arnold and Andrew became very knowledgeable in their particular spheres of interest, which was quite important in terms of their own ability to carry conviction with their political colleagues in the Labour group. They were both people who cared deeply about what the planning service was doing, and felt that planning was potentially important in terms both of its effects on the city and of its capability thereby to improve aspects of people's lives. They were also both people who wanted to know what I and my colleagues thought. They didn't always take that advice, which of course was their privilege, but they always wanted to know what it was. Indeed, as the process of partnership developed they became ever more likely to come and discuss something very early in its life and from first principles, and encouraged me to do the same, which meant that we usually knew where each other stood fairly quickly and could talk on that basis not only about what needed to be done but also about how best to handle the process of doing it.

Exchanges both of information and of more general awareness between us were a very important component of these processes. Arnold and Andrew also worked together very effectively as a pair, which was cemented both by a personal friendship and by the fact that in the later years they both represented the same ward. Arnold was really interested primarily in policy issues, and particularly in the issues surrounding transport and urban sustainability. Andrew, whilst he had been a leading member of the Greater Manchester County Council in the field of transport policy, was an excellent Chair of the Development Control Sub-Committee, who handled the complexity and the detail of development control well and who was genuinely interested in the application of policy to cases. He had accumulated a considerable amount of experience of handling the exercise of public speaking rights in the operation of the Development Control Subcommittee (see Chapter 5), and his ability to deal effectively with this made a large contribution to the success of this as a council initiative which in practice was applied more by the Development Control Subcommittee than by any other part of the council. The complementarity of the pair of them in what they did made the officers' job of relating to the political process in the planning field much easier than it otherwise would have been, because we had no real co-ordinating or mediating job to do in comparison with the situation that many other chief officers reported in their dealings with their chairs and deputy chairs.

We tended to see Arnold Spencer most working days. There was a Chair's office near to mine with the two separated only by a secretarial suite, and this meant that it was very easy for each of us to go in and out of the other's office on an informal basis as necessary. The vast majority of our contacts were of this kind, although there were more formal occasions such as public meetings, committee meetings and chair's briefings where the roles we each played were made more distinct. Andrew Fender was not in as frequently as Arnold Spencer because his paid employment commitments were not as flexible, but he encouraged telephone contact when needed rather than waiting for the next time he was in. Both of them had open access to the staff. I never tried to operate on the basis that all contact should be through me (as I gather some chief officers attempted, although not with much success), simply because I did not believe that such caution or control were either necessary or achievable. I believed that direct access to staff was a good practice in a two-way sense, and if it gave rise to any difficulties in practice then the people concerned were usually mature enough to sort this out for themselves or if they couldn't would simply come and talk to me about it. In addition, both Arnold and Andrew very quickly learned from their own experiences what it made sense in any event to come and talk to the chief officer about, and we always operated on the principle of being as open as possible with each other even when the discussion itself was on a confidential basis. All this builds up quite a strong two-way sense of loyalty, and this meant that when difficulties began to escalate with the political leadership from early 1994 these were experienced both by the executive side of this partnership and by its political side. This probably also meant that this is how we were perceived elsewhere in the council's system; if either side ran into difficulties, the other was bracketed with this whether or not the facts supported this interpretation. Notwithstanding this, because it could certainly be argued that this was a contributory factor in some of the difficulties with the council's political leadership discussed above, I felt that this partnership was an effective and mutually supportive one, which contributed considerably to what the planning service was achieving.

Members of the Planning Committee

As already indicated, membership of the committee came about in a variety of ways, and this was not always a function of a particular interest in planning matters although very often it was. Members of the Planning Committee who had been on it for any period of time tended to develop a good working knowledge as lay people of the limits to action that apply to the planning service, and this combined with their judgements about the relative weight to be attached to issues and representations often made their input not merely decisive (because in a formal sense that was what it was; the committee took a decision) but also constructive.

I have referred above to the tendency in certain cases for committee members to come out of their political group discussions about matters on the agenda with what amounted to a whip on a particular matter. This was not always so. There were often free discussions in committee which led to the emergence of a view about a way forward on a particular matter, and these would sometimes cut across party political divides. In a sense, it was important that this should be so, because in cases where whips were being applied it was all too easy for the party groups on the committee to adopt what was in effect a ritualised opposition to each other's point of view. On several occasions I sat in the committee and heard Liberal Democrat members expressing perfectly sensible points of view, only to find that somebody on the Labour side would immediately say something in opposition to it and party lines would quickly get drawn; sometimes the worst thing that could happen in these terms to a particular point of view was that it got picked up and presented by the minority opposition parties. The ability, sitting alongside the Chair and Deputy Chair, to whisper quietly into their ears when there was a danger that this process might lead the committee into what I thought was a mistake on a particular matter was a very useful facility, although they would always (and understandably) want to be careful in a situation like this in terms of how they handled their own group members.

It would be easy to conclude from a lot of Planning Committee meetings that the only thing that appears to matter is the City Planning Officer's recommendation. Because a high proportion of decisions do follow that recommendation often without any discussion, this line of argument would suggest that it is the views of the officers that are dominant. Clearly, there is something in this. One of the jobs that the officers are there to do is to provide the best advice they can to the committee, starting from the law and practice of town and country planning, the policies of the council as local planning authority, and the particular circumstances of the case. Assuming that the piece of written work that presents the outcome of these analyses is reasonably competent, it would in a sense be surprising (and somewhat disconcerting, in terms of the efficient operation of the system and its ability to relate effectively to its customers) if a high proportion of recommendations were not to be accepted. Members need to feel confidence in their officers if they are to trust them to get on with the wide range of matters that are delegated to them, and in turn officers need to feel confident about both what the policy stances of the committee are and its willingness to apply them consistently in dealing on a daily basis with their customers; and a significant proportion of recommendations proving unacceptable would put these necessary confidences at risk. At the same time, it should be remembered that recommendations on reports do not materialise in a vacuum. Everything that I have already said about working with members illustrates the point that this is both a continuous and a complex process, with a range of interactions taking place all the time. Recommendations on reports come from this background, and in potentially controversial cases are likely to have been discussed politically (notably with the Chair and Deputy Chair of the committee) at formative stages. Thus, what happens in the committee itself is the tip of an iceberg, and the Chair and Deputy Chair are likely to have told their Labour group members something about the iceberg itself in controversial cases at the Labour group meeting before the committee.

Even where the committee and the planning officers end up taking a different view about the outcome these same principles are likely still to hold broadly true. In cases such as this, unless the Chair and Deputy Chair are taken completely by surprise by the strength of feeling within their own Labour group, the report will have been prepared in such a way as to present information and analyses that are capable of supporting other conclusions than those the planning staff have reached, because these things are almost always matters of balance and judgement rather than of absolutes. Members thus may still find the analysis helpful whilst reaching a different view about the balance, and reports need to be seen in this light and not just in terms of what happens to that recommendation. Often in these sorts of circumstances there will have been an informal discussion with the Chair or Deputy Chair beforehand about how a decision based upon a different conclusion might be worded to give it the maximum chance of survival. I never found any difficulty with this. Provided that it was clear that the officers were there to provide the best advice that they could, and were not as a consequence expected to alter their recommendation on the back of a political whim, supplying a written recommendation in one direction and informal guidance about the best way of doing the opposite if that was the political will was not a contradiction in terms. At the end of the day, it is the council that is the local planning authority, not just its paid officers; and since elected members are properly the ultimate arbiters of what the council's position on any matter actually is, their right to take a different view in all the circumstances needs to be respected and not elevated into an unnecessary problem. There can be some difficulties anyway in these sorts of circumstances; for example, a written report on a planning application recommending one course of action will undoubtedly be prayed-in-aid at appeal if the council has done something different and thereby provoked the appeal. This can be potential costs-award territory if it could be shown that the council had behaved quixotically, but to keep this in perspective the council only had three adverse costs awards up to mid-1995 in the entire life of the costs-awards system (all of which were of this kind) and it won far more appeals than this where members had not taken their officers' advice. What is very important is that these sorts of difficulties are accepted on all sides as part and parcel of the process, and are not allowed to get in the way of a general level of confidence being established between the Planning Committee and its officers which must be the bedrock of the system.

Members of other committees whose work impinges on that of the Planning Committee

This type of elected member involvement in the planning process is more or less self-explanatory. A large number of other committees of the council either take decisions which impact upon the planning service, or make proposals or requests that call for some form of response from the Planning Committee. Whilst the days of the council being the largest developer in the city (as was the case during the inner-city redevelopment phase) have gone, there is still a sizeable amount of development activity which is controlled or directly influenced by the council, and much of this requires planning permission. As we will see when looking at the airport expansion case study in Chapter 8, this gives rise to a need to try to keep separate the council's planning interests in such matters from its other interests, in order to ensure not only that justice is done but that it also is seen to be done. The same is also true in the quite significant number of cases where the council has a landowning interest. The theory of this is fairly straightforward, but the practice of it in a highly political world can be anything but!

There are perhaps two categories of member behaviour that are of particular interest in these sorts of circumstances:

- The same members are perfectly capable of taking two directly contradictory decisions about the same matter at two different committees in the same cycle. This isn't as quixotic as it may sound. Members tend to take on the character of the committees on which they serve, and so when they are looking at a matter from a Planning Committee standpoint they will do this from a planning perspective taking the planning advice that is in front of them very seriously. When they are looking at that same matter from the perspective of a developing committee, they equally tend to see the desirability of the development from its service delivery perspective. There is indeed an argument which says that this is perfectly proper, and which leaves the final decision to the Policy and Resources Committee and the council in the event of any conflict of views proving incapable of resolution, although usually an attempt to do this will be made first between committee chairs.
- From time to time there are real and major disagreements between committees which represent not merely different perspectives on a problem but altogether more fundamental difficulties. At the end of the day, the resolution of these will have to be through the political process, but the journey towards that resolution can be both turbulent and very public. In particular, this is just as likely to reflect who has the most political weight at any point in time as it is to be a judgement of the rights or wrongs of the issue in question.

Members of the council

For many members of the council, the distinction between a role they might play in this capacity and in any of the other capacities discussed in this chapter would not be a very meaningful one. For some, however, the rights and duties that go with the role of councillor transcend the other functions and produce distinctive concerns to do with the council's overall role as representative of the city and to do with their individual roles as people with particular rights both of access to documentation and access to staff. Members are in a special position in comparison with all the other customers of the planning service precisely because they are members of the council, and this means that (if they wish) they have a right to take up and follow through a particular issue or matter without needing to rely on any of the other roles. Most don't do this for most of the time, but some do; and it isn't always clear where members are coming from in picking up an issue on this basis. In my experience it is a great mistake to worry too much about what might have triggered a particular member's interest in these terms, because that may well come out at a later stage anyway, and instead officers should concentrate on meeting the request or concern as best the service can at the time that it arises.

In practice, the main way in which an approach of this type often materialised was through a direct but informal approach either to myself or to another senior member of staff often for no other reason than that the member knew the particular officer. This would usually be to get a planning perspective on whatever issue the member wanted to raise, but it might also be to discuss what the best way of handling something within the corporate system might actually be. I always operated on the principle that a member could come and have a confidential discussion with me about any matter that was relevant to my purview as soon as a mutually convenient time could be found, and whilst this wasn't a very frequent phenomenon it was an occasional part of the workload. Very often it would turn out that a member was really after help in thinking through how to take forward something that was concerning them, and its planning dimension might often be incidental to this. Some members, however, had a very strong sense that as members of the council they did have a responsibility to pursue matters of this kind if they felt that the council was not behaving or performing as it should, and they would consequently pursue issues that didn't relate to their ward, their committees, or their known interests. My job, when asked, was to help them with this, without worrying too much about where they were coming from in taking such action.

Members as protagonists of individual policies or views

Many members of the council have some quite strongly held views about some matters. Indeed, it is often this that has marked them out from the mass of the electorate, and caused them to become councillors in the first place. There is usually, although not inevitably, a relationship between a member's interests in these terms and the committees on which that member sits. Almost always, where a member is known to be particularly interested in certain policy areas, this interest will be accompanied by a series of bilateral relationships with groups in the city that are also concerned with these matters. Such groups often seek to target members who are known to be interested in particular areas, because for them this is an effective way of finding people within the council who may be persuaded to act as advocate in particular instances. For the member, these sorts of relationships bring with them both information (and sometimes this is information that is not available through the council's officer structure) and the prospect of generating some political support if groups are active in or connected with party political processes.

The fact that many elected members have strongly held views about many things always needs to be remembered when member–officer relationships are being discussed. The traditional model that presents the elected member as a passive layperson and the professional officer as the technical expert has many faults with it, and this is certainly one. In their fields of interest, some members can and do build up an impressive pool of knowledge and understanding and a wide range of contacts, and are likely as a consequence to be very positive contributors to policy-formulation processes. This can mean that discussions with elected members about individual issues can take place at very variable levels of understanding, but at the upper end of this range member knowledge of itself can be a significant resource. How well this is used within the political process can depend upon quite a range of issues, including how well in that member is with his or her own party's leadership at the time; but the individual member who can give an officer a hard time about a particular issue because of that member's basic knowledge and understanding is a phenomenon that most chief and senior officers do come across from time to time.

Elected members as individuals

Finally, it almost goes without saying that all 99 members of the council are individuals in their own right, and sometimes what they do and how they do it in performing their functions as councillor is a function of this factor rather than of anything else. Indeed, the quirkiness of the behaviour of individual councillors from time to time can be one of the more idiosyncratic components of the life of the council, and it is also one of the aspects in which there can be considerable media interest.

Conclusions

It is important to reiterate the point made in the introduction to this chapter, to the effect that dividing the ways members impact upon the planning process up into nine separate roles is an analytical construct. At any one point in time, members can be performing some or many of these roles simultaneously, but they would not pause to see themselves in this light. Equally, information, understanding and contacts obtained via one role would be regarded as being portable to any or all of the others; members do not build artificial barriers in their minds in these terms. None the less, it is probably more helpful to think of the roles of members in terms of how they actually impact upon the planning process than it is to take a simple and undifferentiated view of their functions. In particular, this analysis perhaps serves to demonstrate the wide range of perspectives that members can take of the planning process when seeking to influence or use it, and if local government officers (and particularly senior officers) are to be effective at dealing with this very important part of the job they too need to handle these relationships with an equally sophisticated understanding of where members may be coming from.

This way of looking at the various roles that elected members actually play in relation to the planning process ought to help to develop a more sophisticated view of officer–member relations than that implied by generalisations to the effect that they are 'good' or 'poor'. What this review has demonstrated is that it is possible for relationships to take both these forms simultaneously, according to which roles are being played by elected members; and no doubt they would retort that this is also affected by whatever roles are being played by officers, since it is clear that elected members do not see officers in a unidimensional way. Officers can find that the natural difficulties that would arise for them in any event in trying to relate to the performance of these nine sometimes conflicting roles can get powerfully reinforced by relationship difficulties that are internal to the political process, as the story of the planning process and of individual planners getting caught in the growing animosity between Graham Stringer and Arnold Spencer vividly illustrates. It can be virtually impossible to maintain the stances of the proper professional officer discussed in the conclusions to Chapter 2 in this situation, especially when whatever is done or said is regarded as either a weapon or a potential blow in a political war which will eventually produce a winner and a loser. The individual planner is also likely to find this kind of problem especially difficult to deal with when it involves the conduct of at least some aspects of this war via the local media, because the planner does not in practice as an employee of the council have equal rights of reply to those assumed by politicians in this kind of situation. None the less, whilst not being at all typical of most relationships with elected members for most of the time, this is part of the range of problems with which the planner has to try to cope, and it can colour most of the others because of its very high visibility.

I hope, therefore, that readers have found this chapter helpful in trying to move away from the undifferentiated (and sometimes very cynical) views that are often expressed about the work of local councillors. On the whole, I would say from experience that most councillors are atypical of the population at large primarily in the sense that they are much more interested in politics than is the average man or woman in the street; put another way, if 99 Mancunians are councillors at any one point in time approximately 439,901 aren't, and probably only 200 or so of those (and this figure may be generous if some of the stories local parties tell about their difficulties in assembling candidate lists are true) would aspire to take the place of the 99. By any test, therefore, this interest in local politics is a distinguishing feature. But apart from this, councillors are in many ways a cross-section of life in the city, and if their interest in politics is what brings them into the process in the first place it is often joined as an element that sustains them in that process by a genuine interest in and concern for the quality of life in the city. This is much more common, for example, than is a driving political ambition, and it is probably also a characteristic that planning staff find easy to relate to. Perhaps this is easy to understand, given that the ethos of planning is that it is about improving the places we all function in and thereby improving the quality of people's lives. What it does mean is that the views of elected members can very quickly come to be respected not merely because they constitute part of the formal decision-taking structure of the council but also because they often come from people who are very knowledgeable about and care very passionately about their local communities. At its best, the combination of these perspectives and concerns and the professional knowledge and skill of staff can be both powerful and effective, and the two in partnership represent the positive face of local government. But it is a face with many features, and hopefully this chapter has thrown some light on some of these.

PART II

THE MAIN TOOLS

4

Approaches to development plan-making

Introduction

This chapter looks at development plan-making in Manchester during my own first-hand experiences of it, from a period in the 1980s when the focus was mainly on developing informal styles of local planning, to the early 1990s when the emphasis was on the preparation of the first Manchester Unitary Development Plan. I have looked elsewhere at the history of development plan-making in the city going back to 1945 (Kitchen, 1996e), and as well as referring to that paper readers who want to pursue this fascinating story can look both at the original documents and at the views of other commentators.

The main original documents are the postwar Reconstruction Plan prepared by Roland Nicholas (Nicholas, 1945), who was then City Engineer and Surveyor, and who was responsible for many other functions as well; the city's 1947 Act Development Plan (Manchester City Council, 1961), which was strongly grounded in the Nicholas Plan and which took 10 years to get from submission to the Secretary of State to incomplete approval, demonstrating many of the criticisms of the 1947 Act system subsequently to be made by the Planning Advisory Group (1965); the City Centre Map (Manchester City Council, 1967), which sought to deal with the part of the city where the proposals of the Development Plan had not been approved by preparing a non-statutory plan for the city centre along the lines then advocated by the Ministry of Housing and Local Government; the attempt by the North West Joint Planning Team (1974) to assemble a regional strategy putting more attention on the Manchester and Liverpool conurbations; and the documentation associated with the Greater Manchester Structure Plan (Greater Manchester Council, 1979), including the report of its Examination in Public Panel (Vandermeer, Mill and Morrison, 1980).

The main commentaries are provided by Wannop (1995, pp. 136–50), who looks at the regional context throughout the twentieth century; Hall and his colleagues (Hall *et al.*, 1973, pp. 566–610), who look at the lack of success of the planning system at that time in coping with the economic and social changes adversely affecting the northwest region's two major conurbations; Turner (1967, pp. 60–76), who took a notably gloomy view both of the state that Manchester was in at that time and of the attempts made to date through the planning system to tackle these problems; and Kavanagh (1971), commenting on the efforts of the North West Economic Planning Council. As far as the GMC Structure Plan is concerned, the main specific commentaries are by Fenton (the GMC officer who led the county Structure Plan team; see Cross and Bristow, 1983, pp. 28–61) and by Healey *et al.* (1985a; 1985b). More generally, Caulfield and Schultz (1989, pp. 13–15) look at the common problems of structure planning, largely following the contemporary criticisms of it advanced by the Thatcher government, and Barlow (1991, pp. 145–47) looks at the reasons why the metropolitan county councils such as the GMC were abolished in 1986. Finally, Ankers, Kaiserman and Shepley (1979) have produced a well-known parody of county–district relationships, which illustrates in comic form some of the real difficulties that were experienced between the two tiers of planning authorities in the conurbations in the 1974/86 period. Apart from these direct pieces of commentary, there are also parallel studies of the redevelopment versus refurbishment debate at different stages of its life in Liverpool (Muchnick, 1970) and Birmingham (Paris and Blackaby, 1979), and the development of major national policies such as the emergence of public participation in planning as a statutory right, grounded in the work of the Skeffington Committee (Skeffington *et al.*, 1969).

A chapter about development plan-making is at risk of treating the achievement of a development plan as an end in itself. My view of this is that the development plan is one of two major statutory tools that the planner possesses, the other being development control, to contribute towards the continuous objective of making our towns and cities better places within which people can carry out their lives. The development plan is thus one important means to this end, but it is not the end itself. Indeed, I would also argue that the statutory mode of development plan-making can at times get in the way of the achievement of this objective, by virtue of some of the procedural and the content issues that surround it, and I would claim that the history of development plan-making activities in Manchester supports this view. My tentative conclusions from the historical record (Kitchen, 1996e, pp. 346) were as follows:

- The city had experienced in the 50 postwar years three periods of intensive formal plan-making, each lasting approximately 5–6 years, with much longer informal interludes in-between.
- The informal periods were mainly dominated by the generation of non-statutory documents and the decay of the existing statutory documents.
- The output of the formal phases was more valuable at the level of broad strategy than it was at the level of detail needed to deal, for example, with the redevelopment process in a locality.

- Statutory processes were very time-consuming, and statutory documents began to decay during the processes because the world did not stand still whilst due process was being observed.
- The predictive capacity of planning documents was very variable, with a particular weakness in the field of anticipating changing economic circumstances.

Plan-making in Manchester in the 1970s and 1980s

Brian Parnell, my predecessor as City Planning Officer of Manchester, had spent many of his formative years handling the inner-city redevelopment process, and had come to value through this experience the flexibility of informal strategies and plans whilst recognising the limits of the statutory development plan. These were not two polar opposites, and some statutory local planning was undertaken during his period of office, but his view was that the case for a statutory approach needed to be made in each particular instance; and if it wasn't clear-cut then he believed that informal and non-statutory approaches should prevail. Perhaps the main components of this were as follows:

- Doubts about the need for statutory plans when the City Council reputedly owned at one stage over 65% of the land surface of the city and in some areas virtually all of it. Brian believed that a combination of landlord controls and ordinary planning controls should be more than sufficient in these situations to enable the City Council to achieve what it wanted.
- Doubts about the worth of the time that the statutory process took in terms of the extra benefits it generated as compared with the speed and flexibility of non-statutory approaches.
- Worries about the inflexibility of 'end-state' planning when incorporated into statutory documents. An example was the concept of 'walkways in the sky', which was developed in Manchester (and many other cities) during the 1960s partly in response to the Buchanan Report (Buchanan, 1963). This depended upon sets of buildings being reconstructed to provide linked upper-level pedestrian walkways, but until enough sites were redeveloped on this basis and joined together the walkways that were constructed as parts of individual buildings were of neither use nor ornament. Several individual redevelopments of this kind were undertaken in both the Higher Education Precinct and the city centre areas, but the linked systems of which they were intended to be part never materialised and they increasingly looked like embarrassments rather than the brave new world of which they might have been part had the idea materialised in all its glory.
- A desire to see the planning process behaving differently from in the past, partly in response to changing community and political views about the inner-city redevelopment process and partly because of a genuine belief that more community contact and involvement should be encouraged without as

many fixed points (which could so easily be interpreted as 'we know best') as had been the case in the past.

What all this actually meant in practice was an approach to development planning in Manchester City Council in the 1970s and 1980s, which had four components to it:

1) A willingness to let the 1961 Development Plan fade away into irrelevance. It had performed its primary functions in support of the redevelopment process, and the view was that actively seeking to bring it up to date by various means would be time-consuming and largely counterproductive. Whilst recognising that it had a continuing life in a statutory sense, which included in the realms of development control, the general view was that the importance of this should be minimised and that the 1961 plan should not be seen as a constraint on what the city wanted to do.
2) Active participation in the structure planning process to try to ensure that it remained at the strategic level and didn't introduce new and unwanted constraints on decisions at the local level. In effect what this meant was that the structure plan was seen as having the potential to go beyond what a strategic document ought to be, and thereby to replicate in this new form some of the problems of the styles of planning from which the city was trying to move away. This was a component in the arguments about trying to scale down the first Greater Manchester Structure Plan, which were developed on a joint basis by the 10 district councils in 1979, which I participated in on my arrival in Manchester in that year and in which the City Council played a leading role. This resulted in representations against roughly 60 of the 160 policies in the structure plan being presented on behalf of all the districts at the examination in public, 50 of which were accepted by the EIP Panel (Vandermeer, Mill and Morrison, 1980, para. 1.2). What this did not mean, however, was the City Council saw no role for the structure plan; it wanted a strategic level of planning support, but it did not want its local discretion to be fettered unnecessarily.
3) A limited role for statutory local plans, only where this was justified by specific circumstances. The statutory local plans on which the City Council did embark were

   - the City Centre Local Plan, because the scale and complexity of the city centre and the limited amount of City Council landownership meant that the council needed to have more in its armoury that in most parts of the city;
   - the Ringway Local Plan, to help to manage the continuing expansion of Manchester Airport; and
   - the Beswick Local Plan, which was almost a symbol of a new age and a new approach because Beswick was the first of the areas of system-built high-rise housing from the 1960s/1970s to be demolished.

   In addition, there was joint working (with the GMC in the lead) through the River Valley Committees in the Mersey and the Medlock Valleys which eventually produced statutory local plans, and the GMC itself produced a Green Belt Local Plan and initiated a Minerals Local Plan (although this was not completed and then adopted until well after the GMC had disappeared).

Thus there was some statutory local plan coverage, although the vast majority of the city was without it, and in the City Council's view was no worse for this.

4) Informal approaches at local levels, increasingly working with local communities. This was fertile territory for the development of integrated area teamworking, and also for the emergence of thinking about client/customer approaches to the planning service that was described in Chapter 2. Perhaps more than anything else, however, it paved the way for the development of much more pragmatic approaches to planning than the city had previously seen. The apotheosis of this view in the mid/late 1980s was the emergence of a model which sought to combine the publication of a high level of information at ward level about the social, economic and physical circumstances of the area with the publication of ward planning statements in leaflet style which went through every door in a particular ward. The intention behind these was to give a straightforward account of the planning context, circumstances and issues facing each ward, then to set down the council's current thinking about these matters either in terms of policies or (where specific views existed) at site scale, and then to invite public response and dialogue on the basis that if such activity could show that change was necessary and desirable these documents were easy to change. The ward was chosen as a level of activity not because wards were natural planning areas (the debate about what constitutes a 'natural' planning area is probably endless anyway) but because political and quite a lot of community processes related readily to it; and because at 33 wards in a City of Manchester's size (or an average of between 13,000 and 13,500 people per ward) it provided a relatively local basis for this work. Just before this round of work came to an end as a result of the combination of staff reductions and the impending arrival of the UDP process, 18 of the 33 wards in the city had been covered by this approach, and a standard format was beginning to emerge both from experience and from customer feedback to replace the various experiments that had been tried at the beginning of this initiative. The Manchester approach in the early 1980s, just before the ward planning statements initiative was attempted, has been written up in Fudge and Healey (1984), and its main components are summarised in Healey *et al.* (1985b, pp. 39–41).

The Manchester Unitary Development Plan – background

Many of us were not sure that the government of the day was wholly clear about what it was doing when, as part of the process of abolishing the metropolitan county councils, it introduced the concept of the unitary development plan. Perhaps this is reflected in the relatively slow rate of introduction of the concept across the affected conurbations in the form of the promotion of commencement orders by the Secretary of State for the Environment; in the case of Greater Manchester, the 10 districts didn't receive their commencement orders until more than three and a half years after the disappearance of the GMC. Commencement orders were promoted at different rates across the country as a whole, however, and it appeared from the perspective of someone on the receiving end of this process that one of the component elements of this was a perception in the DoE of the readiness of authorities in terms of the existence of a strategic framework at conurbation scale to guide UDP work. In the Greater Manchester case, this appeared to be seen as doubts about the ability of the 10 districts to co-operate on strategic matters in the absence of the GMC. If this was so, the Association of Greater Manchester Authorities' advice to the Secretary of State about what strategic guidance he ought to give the districts (Association of Greater Manchester Authorities, 1988) must have given some encouragement about the possibility of effective joint action. This document was largely drafted in Manchester City Planning Department, deriving in part from Manchester's quite long-standing role as Secretary of the Greater Manchester Planning Officers Group (which I continued on Brian Parnell's retirement in 1989) which oversaw the preparation of the document on AGMA's behalf.

By comparison, the strategic guidance that was ultimately issued in October 1989 (Department of the Environment, 1989), to be followed by a commencement order in December 1989, was both much briefer and rather less precise that the joint advice that had been given to the Secretary of State (see Wannop 1995, pp. 146–48). Whatever the shortcomings of strategic guidance, however, its arrival ushered in a new world of plan-making, where to a degree debates about formal and informal, statutory and non-statutory had been rendered irrelevant by this instruction to prepare a UDP in this context.

In the light of what follows about the UDP preparation process, it is instructive to recall what was actually said by the DoE at the time about these matters. The covering letter to the strategic guidance document from the Regional Director said the following about this:

> The next step is the preparation of the Unitary Development Plans. To be most effective, each should be clear and brief, produced to a brisk timetable and kept up-to-date.
>
> There has been substantial consultation during the preparation of Strategic Guidance. Further consultation looking at more local issues should be undertaken as part of plan preparation. Generally within Greater Manchester the Secretary of State expects that it should be possible to have draft UDPs on deposit within two years of commencement.

One of the implicit messages in this, which chimed well with our thinking in Manchester, was that in carrying out this function it should not be necessary to reinvent the wheel. Not only did this make sense in terms of the time and other resources that the whole process was potentially capable of consuming but it also fitted in with a desire to build on the foundations that already existed in the city by virtue of both the statutory and the non-statutory planning work that had already been done and the information and understandings drawn from ongoing relations between the Planning Department and all its customers. From this perspective, there were four primary inputs into the UDP process in Manchester:

- The Secretary of State's strategic planning guidance, because we had to have regard to it. That having been said, for the most part the document dealt in (not unhelpful) broad generalities, and had very little in it that was specific. Indeed, its only really precise components were an invitation to the 10 districts to co-operate in the preparation of a portfolio of major high amenity sites for high-technology industry and a specific allocation to each district of a target figure for new dwellings between 1986 and 2001 to a county control total of 60,000.
- The existing development plan framework, consisting of the Greater Manchester County Structure Plan, the adopted local plans in Manchester and the 1961 Development Plan (although in this latter case only residually so).
- The accumulated experiences of and outputs from the informal and non-statutory local planning work that had increasingly become the dominant element in planning work in Manchester in the 1970s and the 1980s, which has been described above.
- The bank of knowledge and understanding that had emerged from developing relationships with the customers of the planning service, described in Chapter 2.

One of the most important management decisions taken about the UDP process in Manchester from the outset was that the existing levels of information and understanding represented a broadly satisfactory basis for plan preparation. No 'report of survey' was undertaken in any conventional sense of this phrase, and no major new pieces of information gathering were commissioned. Rather, what we decided to do was to recognise that we had three broad sets of staff resources that we could draw upon, and that we would make more progress by getting the understandings that these people had on paper as quickly as possible so that we could begin to generate reactions from the customers of the service than we would by an intensive and possibly extensive period of navel-gazing. In its essence, therefore, this thinking was very similar to the thinking that had driven the work on ward-planning statement described above. The three main internal sources were as follows:

- The Strategy Group, to take and maintain a strategic overview both of what we were trying to achieve and of the process itself, so that the group became the primary 'holder of the pen' to ensure as far as necessary and desirable a consistent end-product.
- The department's area teams, to input local knowledge and experience and to prepare first drafts of what were to become the area-based sections that constituted Part 2 of the plan.

- The department's specialist groups, to contribute throughout the plan from their perspectives but in particular to input to the policy-based sections that constituted Part 1 of the plan.

The internal working method that was used was essentially one that emphasised structured debate. I chaired a series of meetings around particular themes, where the usual pattern was that a draft of possible policy stances was prepared by the appropriate people (it could have been any of the above three sets depending upon the subject matter), circulated for a meeting at which each of those groups was represented, and then subsequently amended in the light of what had been agreed as a result of the discussion. At the end of the day, I suppose I had more votes than anyone else, although this was used sparingly as against the desire to identify broadly internally consistent majority positions. Because we were very keen that our broad policies should be tested out, Part 1 and Part 2 were not tackled in sequence but by an iterative process. Strategy Group took the responsibility for chasing up the multiple drafts that were on the stocks at any one point in time, and then for identifying any major internal inconsistencies which were arising and therefore needed to be discussed at the general meetings. What we did not do was keep putting drafts back to those meetings for endless repetitions of the same discussions; the principle that was followed was that people could be trusted to reflect the outcome of discussions, with the editorial and progress-chasing functions of the Strategy Group and the final say-so when necessary of senior management (either the assistant city planning officers or if necessary myself) acting as the checks and balances. The products of this potentially somewhat anarchic process were bounced off the Chair and Deputy Chair of the Planning Committee at appropriate times, and this kind of political steer was very useful. So also was the fact that targets were kept in mind all the time, in the light of decisions about public consultation and involvement which are described in more detail below.

We were much helped in all this by two decisions by elected members about the nature of the process itself:

- They wanted us to follow a consultative style, as the City Council was increasingly seeking to do across the board, which went beyond the statutory minimum and sought to ensure both that the draft plan would be well grounded in the views of the customers of the service and that when members came to have to make decisions about the draft plan they could do so with a very good level of understanding of what public views actually were.
- Because it was acknowledged that people would not be likely to think that the council was very open to representations seeking to change the contents of a draft plan if it had already approved that plan as a basis for public consultation, it was decided that the council formally would not seek to clear a consultation draft at all but would only look at documentation later in the process to determine what should become the deposit version of the UDP when consultation had already taken place and when its results could

be reported. In giving me and my staff this amount of delegated responsibility, it was both expected and understood that we would work very closely with the Chair and Deputy Chair of the Planning Committee as this process unfolded. From discussions with colleagues up and down the country it is clear that this was an unusual decision, but it enabled us to approach the consultative process in such a way as to be able to demonstrate to people that this was not a *fait accompli*.

Public consultation and involvement

The upshot of all these considerations was a decision to try to embark upon an element of public consultation on issues and choices, in advance of the statutory requirement to consult either on the matters to be contained in the draft plan (as it was at the outset of the whole process) or more minimally on the draft plan itself (as it became as the process unfolded). The vehicle chosen for this was a free newspaper to go through every door in the city (although it subsequently became clear that some difficulties had been experienced in achieving delivery on this basis, and as a consequence less than 100 per cent coverage was achieved). The purpose behind this was not so much the belief that we were likely to be able to generate a massive response, but rather the view that by setting down our perceptions as openly as possible at an early stage of the process we could encourage a dialogue with people who wanted to contribute on various matters, including those who might feel in the light of what they had read that we had got something seriously wrong and might therefore be provoked to say so. It was also useful as an early focus for work to have something like this to aim for. Box 4.1 demonstrates the changing contents of the three editions of the free newspaper that were issued.

This process was one of the national exemplars of good practice in the local delivery of planning services subsequently picked out in a major national survey of local planning authorities (Spawforth and Rankin, 1995). That survey (*ibid*., p. 19) commented thus on what we were attempting here:

> There are few things less exciting than a UDP for the average citizen. Three staged issues of 'City Planning News' tried to ensure that everyone on first reading could see something relevant to them and that the relationships between components of urban life were clear.
>
> Assistant City Planning Officer David Kaiserman says people expect planners to be able to express succinctly what they are there for. 'Because of the many varied pressures on planning, that is not easy, but it is essential' he says. Like the UDP, the newspaper had to be accurate, comprehensive and credible, a challenge Manchester's planners rose to.

It is always encouraging to have one's efforts commended by one's professional peers in this way. What is harder to assess is whether or not the processes that we went through ensured as good a level of customer input to the plan and as high a level of customer satisfaction with it as we could have

Box 4.1 Components of UDP consultation free newspapers

City Planning News in the form of issue number 1 was published in December 1990. It was a 16-page newspaper, consisting of 5 pages of general material on key issues across the city as a whole, 10 pages of commentary on what all these might mean and also on more local issues in each of 16 geographical subareas of the city, and a back page which sought to convey the essence of what we were saying in seven ethnic minority languages (Chinese, Vietnamese, Urdu, Hindi, Somali, Gujerati and Bengali).

This was followed up at draft plan stage by issue number 2 of *City Planning News*, dated September 1991, where the delivery process to households was more successful than had been the case with issue number 1. This second issue summarised what was being said in the draft plan down to 16 newspaper pages, broken down as follows:

- Three pages on general policy issues.
- One page on how the policies and proposals might start to affect particular groups of people in the city (children and young families, elderly people, disabled people, young adults, women, ethnic minorities, gay men and lesbians, low-income groups).
- Ten pages on what this might all mean in the 16 geographical subareas of the city.
- A ½ page on some specific development control policies.
- A ½ page on arrangements for a touring exhibition about the draft plan at 40 different venues in the city between 23 September and 30 October 1991.
- One page summarising the essence of this in eight ethnic minority languages, with Punjabi having been added to the above list.

A third, much briefer, edition of *City Planning News* was prepared as a four-page insert into the council's *Civic Review*, which goes into every home in the city. This edition, dated April 1992, was simply designed to tell people that the City Council had now taken decisions about what was to be the deposit version of the plan, what the process was from that point on, what the major areas of public response to the consultation on the draft had been, and what the council had done in responding to these points. Because the available space this time was much less than for editions 1 and 2, no attempt was made to include an ethnic minority languages component in edition 3.

hoped to achieve. This is probably not readily capable of assessment, especially since on the whole in exercises of this kind people appear to be more prone to make comments on what they do not like than on what they support. It is dangerous, on the other hand, to assume that the absence of comments betokens general satisfaction, especially since despite our efforts we were still coming across people later on in the process who were affected by specific policies or proposals who said that they had not been aware of anything to do with the UDP up to that point. What we probably can say with some confidence is that what we did here did go well beyond what we needed to do to meet statutory requirements, that it did represent very genuine attempts to communicate with people in ways that they were likely to be familiar with, and that we did pick up and use fully the comments that we did get. For the record, the numerical position at consultation draft and at deposit stages was as follows:

- At consultation draft stage, 355 comments were received.
- At deposit stage, 1706 representations were made, including about 340 in support of particular policies or proposals. After negotiations, about 1250 objections stood for consideration by the UDP inquiry inspector, including 425 from the Department of the Environment. This latter cluster is discussed in more detail below.

No equivalent calculation can be done in respect of the responses at the earlier issues and choice stage by its very nature, but in very rough terms it would certainly have been somewhat less than the level of responses received at consultation draft stage.

Box 4.2 UDP timetable

1) Commencement order – December 1989.
2) Consultation on issues and choices – December 1990. Running total 12 months.
3) Consultation on draft plan – September/October 1991. Running total 21/22 months.
4) Decision by the council on the content of the deposit version – March 1992. Running total 27 months.
5) Deposit – October/November 1992. Running total 34/35 months.
6) Public inquiry – July to December 1993 (but it did not sit continuously). Running total 43/48 months.
7) Receipt of Inquiry Inspector's report – August 1994. Running total 56 months.
8) Deposit of proposed modifications in response to the inspector's report – February/March 1995. Running total 62/63 months.
9) Adoption – July 1995. Running total 67 months.

Timetable

The major stages in the process are set out in Box 4.2. It can be seen that it took 5 years and 7 months to get from commencement order to adoption, which was the fastest rate of progress of any of the 10 Greater Manchester districts. Of this time, 2 years and 10 months (fractionally over 50 per cent) was taken up with getting to the deposit stage, and the balance of 2 years and 9 months was taken up with moving through the statutory processes from deposit to adoption. The fundamental difference between these two halves, of course, is that whilst the City Council's behaviour as local planning authority tends to structure the first half of this process, the second half is largely out of its hands. Indeed, almost as a reflection of this distinction, whilst I personally played a full part in the process up to the point where the council decided on what the contents of the deposit version should be (and more or less immediately afterwards went to act as the Council's Chief Executive for a while), after that point my role was a relatively small one. In particular, we decided for good practical reasons that David Kaiserman (my deputy as City Planning Officer and also Head of the Area Planning Division) would play the time-consuming role of lead officer at the UDP inquiry, and like all inquiry processes once a decision like that is taken the team so chosen (including external legal advice) have to be left to get on with it in the ways they think best. Whilst we did not manage to meet the DoE's target of getting to deposit within two years of receiving the commencement order, looking back on the process it is difficult to see how we could have saved very much time given the decisions that were taken about the nature and the level of public involvement we were seeking to secure. It is of course possible to argue that this (which goes well beyond the statutory minimum) could have been reduced, but as a counterargument it could also be claimed that time saved in the earlier stages by reducing public involvement merely to that necessary to meet the requirements of statute might have produced more objections later and therefore more time lost in these later stages. We will never know, of course, but some credence might be given to the claim that this was overall an efficient and an effective process by a recognition of the fact that Manchester as the largest of the 10 Greater Manchester districts was also the quickest to go through the entire process.

The Manchester experience in terms of speed through the process can be compared with the work reported in the early 1980s by Bruton, Crispin and Fidler (1982). They looked at a range of local plans, and concluded that the average time taken from commencement to adoption was 52 months. This compares with the 67 months taken by the Manchester UDP. The period up to deposit in the national sample took 36 months (69 per cent of the total) as compared with 34/35 months for the Manchester UDP, and the phase from deposit to adoption in the national sample took 16 months (31 per cent of the total) as compared with 32/33 months for the Manchester UDP. Thus, the Manchester experience up to deposit was very similar to the national average, and the whole of the difference between the two situations (because overall the Manchester UDP took 15 months longer) is to be found in the processes that took place postdeposit. This made the difference between these two phases being split roughly two-thirds: one-third in the national sample (*ibid.*) as against roughly half: half in the Manchester case. Looking at this in more detail, most of this difference is explained by the length and complexity of the inquiry process and consequently the time taken to receive an inspector's report in the case of the Manchester UDP. Bruton, Crispin and Fidler's average for these phases was 9.5 months, whereas they took approximately 21 months with the Manchester UDP. This may be a function of the greater complexity of a UDP as compared with an average local plan, and also of the fact that the strategic component was largely settled in this latter case, because Bruton, Crispin and Fidler were reporting on local plans in a two-tier system where principles had already been determined by the structure plan. It may also be simply that in the intervening period the propensity to object had grown, a process which Section 54A of the Planning Act is likely to encourage further.

The nature of the UDP document

As far as substantive content is concerned, the draft Manchester Plan (Manchester City Council, 1992) took as its basis two packages of broad objectives each relating to a major theme. These two themes are:

- improving the city as a place to live, work and visit; and
- revitalising the local economy.

Our first attempt at this, which found expression in the December 1990 consultation newspaper, did not take precisely this form, but instead proposed three themes:

- Improving the city as a place to live.
- Revitalising the local economy.
- Enhancing the city's role as the regional capital.

The change to a simpler bipartite approach to major themes came as a result of the internal and external debates provoked by the work at issues and choices stage, essentially through seeing matters to do with the regional capital as being integral to rather than separate from the other two themes.

Part 1 of the plan then sets out the strategic framework for the city by giving expression to the major themes and their packages of objectives at city-wide level through a fairly standard set of topic-based chapters. Part 2 of the plan, which perhaps is not so standardised, divides the city into 17 subareas, as compared with the 16 that had featured in the consultation free newspapers. The difference is that Hulme in the deposit draft version was treated as a free-standing area because of the council's success in securing City Challenge resources for its regeneration, whereas previously it had been part of a larger area covering both Hulme and Moss Side. The basis for the definition of these areas was that they were made up of clusters of the council's neighbourhood office areas, which had been defined corporately in the late 1980s so as to provide a common basis across departments for the delivery of local services and initially in the hope that the council could work over time towards having an office for each such area. On average, there were about three neighbourhood office areas to each UDP area, with the clustering undertaken on the basis that the chosen boundaries made some sort of community and planning sense. We were very keen at this level of presentation of the UDP that it was as user-friendly as was possible, and so in each of these areas we tried to ensure that what was said about the area was as comprehensive as possible and did not rely on extensive crossreferencing to Part 1 (although there was, of course, a health warning in the introductory material about the need for Parts 1 and 2 to be read together). Thus, each of the 17 area packages in Part 2 opened with a general policy about the key intentions towards the area as a whole, usually expressed as a set of objectives for the area, accompanied by a fairly full set of reasons for this general approach; and then the more specific policies for sites or parts of the area were set out in an order which follows on from and gives effect to the objectives for that area.

This approach to the creation of Part 2 of the UDP (which also included a separate section on development control policies) was reflected in decisions on the format of the document. We wanted to achieve a document that made the reading of text and of related plans as easy as possible, given that many of the customers of the UDP process were not necessarily very familiar with reading plans or with handling this kind of complex relationship between text and plan. Consequently, we decided to produce a large document in landscape format that people sometimes found a little clumsy to handle at first, but that generated quite a positive response once people had got used to its unusual format precisely because relating text and plans was easier than was often the case with planning documents. There is often a tension between these various elements, which the UDP Inquiry Inspector (Midgley, 1994, para. 17) summarised as follows:

> The City are to be commended on the attractive appearance and presentation of the Plan. The large format enables the various sections of the Proposals Map to cover extensive areas, facilitating the use of the Map for reference purposes. Conversely the large format makes the document rather cumbersome, and the interspersion of A4 pages of text with the maps reduces the ease with which the written material can be speedily located.

The response to this with the published adopted version (Manchester City Council, 1995e) was to retain the basic approach because its advantages were thought to outweigh its disadvantages, but to improve its internal cross-referencing to help with the point about the speed of location of written material, and to produce it in loose-leaf format so that individual pages could readily be replaced as the plan was updated.

Monitoring and review

One of the most difficult issues that has to be dealt with in a process as time extensive as this is the problem of ensuring that the UDP is kept up to date so that its continuing relevance is secured. If this isn't achieved, the value of the process to the local planning authority is put at risk, and (worst case) the UDP precisely because of its statutory significance could even become a barrier to sensible decision-making in individual instances. An important component of this was the resolution of the City Council in March 1992 that for development control purposes it would from then on rely first and foremost on the version of the UDP that it had just agreed should go on deposit. There was obviously an element of risk in this at appeal, in the sense that a deposit draft UDP would not carry as much weight as would the adopted plan. On the other hand this decision meant that the city would immediately begin to benefit from all the work that had gone into the production of the deposit draft of the UDP, and our hope was that the efforts that we had made at achieving effective public consultation would help to ensure that an inspector on appeal did attribute as much weight as possible to the document notwithstanding its draft status. In the event, this did not turn out to be a problem.

This was also a way of coping with the phenomenon of which we were only too well aware, which was that the plan in a sense would begin to decay immediately a version of it had been settled by the City Council as being the version to proceed through the statutory hoops. Clearly, this would be a very variable phenomenon, and we tried to cope with it (for example) in the major urban regeneration areas by acknowledging that work would continue to be undertaken, that the council would therefore be publishing more detailed strategic development guidance for those areas as that work unfolded, and by concentrating on basic principles in the draft UDP. Hulme (one of these major urban regeneration areas) was very much a case in point, where the draft UDP was relatively skeletal because at the critical point in relation to the drafting of the UDP the City Challenge process for Hulme was still evolving. In that case, on the back of a technical objection to the framing of Hulme policies, we took the opportunity to introduce firmer guidance about Hulme via the UDP inquiry process, because we in consultation with Hulme Regeneration Ltd (which by that time was fully up and running, which was not the case when the UDP was being drafted) were much better placed to do so. None the less, the fact that policy for Hulme was evolving throughout this period proved to be a real problem in terms of the UDP in the Stretford Road case (see Chapter 7), because of the difficulty in a statutory sense of introducing changes during the period between the closure of the UDP inquiry and ultimate adoption of the plan. The essence of this difficulty was that, once adopted, individual bits of the plan could be changed through the statutory process without placing the whole of the rest of the plan in jeopardy, but until that position had been reached there appeared to be no way of introducing changes without setting the whole process (and therefore all the plan's contents) back for a considerable period of time. This illustrates one of the real practical problems that can arise as a result of processes where 50 per cent of the total time is consumed by statutory steps from the deposit stage and beyond.

One of the most important decisions that we took in this regard was that the standard mechanism for keeping the plan up to date once it had been adopted would be via an annual report on development issues and progress. The intention was that this would review these matters in so far as they related to the UDP, and would become the basis for promoting changes when these were demonstrated as being desirable. At the time of writing, this was not a mechanism that had been fully tested because adoption had only just taken place, but the need for it was clearly demonstrated by the fact that in parallel with the statutory steps involved in moving towards adoption a handful of issues had already accumulated where some changes to the UDP needed to be examined. Examples here relate to some quite controversial issues: the generation of a *City Development Guide* (see Chapter 7), progress in the major regeneration areas and the difficulties the City Council had been experiencing in dealing with planning applications for the various types of special needs housing. It could, of course, be argued that the UDP had been at fault in not dealing with these issues properly in the first place, and there may be something in this; but all these examples relate to matters which had become controversial after the UDP inquiry had closed, and thus were really part of the continuous process of adaptation with which the statutory steps are not very good at coping. The effect of all this was that it was proposed to promote the first set of alterations to the UDP only a few months after it had been adopted, and then to try to keep on top of this through the mechanism of the annual report. This would not preclude the possibility that individual alterations could be promoted as the need for them arose, but it ought to make this an exceptional rather than a normal circumstance.

The monitoring and review work that is needed to make a process of this kind successful needs to be undertaken on a continuous basis. It also needs to consist both of formal and of informal elements. By their very nature, formal studies tend to take place at particular points in time and to stand for a period of time. An important example of this, of critical importance to the UDP process, is the need to keep on top of and to publish regular information about land availability in the city. The Planning Department since 1984 had been publishing information about development activity in the city and the total amount of land available to accommodate it, and the report on this published in June 1994 (Manchester City Council Planning Department, 1994) was the tenth in this series. That version assembled information on the basis of the 17 UDP areas that constituted Part 2 of the UDP, and amalgamated this information into four subareas of the city that as near as possible coincided with the area team boundaries that had been the basis used for previous reports, so as to continue with broadly consistent data streams. Clearly, a report like this by its very nature is a valuable input into the UDP monitoring process, but equally and for good practical reasons it needs to be a free-standing document prepared and published to its own timescale which acknowledges its role in relation to the UDP as one (but only as one) of its primary functions. If the UDP annual report were to attempt to encompass between two covers all the relevant information from all the streams of activity such as this, it would become a voluminous document that would undoubtedly be in danger as far as its readers were concerned of failing to see the wood for the trees; so what is needed here is an explicit understanding of how a relationship of this kind is intended to work, rather than either document being seen as a substitute for the other. In addition to these formal studies, however, there is also a need to ensure that an annual report of this kind incorporates the essence of customer understanding and feedback that is emerging in particular from the work of the area teams, and the process of ensuring that this is done in such a way as to make an effective contribution to the report is both a managerial and a drafting challenge.

The DoE's objections to the Manchester UDP

Arguably the most discordant element in the whole UDP preparation process was the role the DoE chose to play at the deposit stage. As I have already said, by the time that the UDP Inquiry Inspector had to give consideration to the objections that remained after negotiations had resulted in agreements about over 100 of them, 425 of the 1,250 (just over one-third) were from the DoE. Very few of these (and on a less charitable analysis, virtually none) were about matters of substantive policy; they were almost all about the ways in which particular policies were expressed or about whether in the DoE's view they amounted to land-use policies, and matters of this ilk. Quite why the DoE felt itself duty bound to do this, not just in Manchester's case but in the cases of the other Greater Manchester UDPs as well, was never explained. Nor were they willing, despite advising that this process applied to everyone else, to appear at the UDP inquiry and have their objections debated; we had to accept them as they stood, give whatever consideration to them that we thought was appropriate, and acknowledge in so doing that the threat of the Secretary of State's intervention in the later stages of the process was at least implicit if we had not in the view of the DoE taken adequate account of their views. Our perspective on this was that on the whole it was a trifling contribution. It would have been much easier for us to understand, and to respond constructively to, objections about substantive matters of policy in which the DoE had an acknowledged interest than it was to respond positively to what appeared to us to be unnecessary and often not very helpful grannying.

The view we ended up taking, however, after much debate, was that there was no point in courting difficulties here when it didn't really matter. So, we agreed to accept DoE objections of this kind when it didn't actually make very much difference one way or the other, and only stand our ground when we either didn't agree with the DoE and felt that their intervention changed in some way the essence of what we were trying to achieve with a particular draft policy, or when it would have involved putting at risk a policy that had been the subject of positive representations via the UDP consultative process. This strategy proved to be an appropriate one, because the Inquiry Inspector was broadly supportive of what we were trying to do here and (no doubt as a result) the DoE did not chose to intervene in the later stages of the process leading to the formal adoption of the plan. None the less, this part of the story raises questions about what the role of central government in the UDP process ought to be, and how such a role relates to the time consumed by the whole process; and I return to this below.

Emerging regional planning guidance

Finally, work on the Manchester UDP could potentially have been disrupted in its latter stages by the emergence of draft regional planning guidance. There is always an argument in plan preparation exercises about top-down as against bottom-up approaches, and there may well be things to be said in an ideal world about the value of having regional planning guidance in place before embarking on the preparation of a UDP for the region's core and 'capital' city. But in the real world, things didn't work in that order. What actually happened was that we received the Secretary of State's strategic guidance covering the whole of Greater Manchester and then very shortly afterwards a commencement order in 1989, and the process of assembling regional planning guidance was not embarked upon until much later when we were very well into the UDP preparation process. At the Secretary of State's invitation, the North West Regional Association submitted advice in March 1994 about what it wanted to see incorporated in his Regional Planning Guidance (North West Regional Association, 1994). Inevitably this was itself the result of some compromise within the Association in order to get agreement on a document, and Manchester had played a full part in this process as a member of the Association's Regional Planning Guidance Subgroup that had actually done the drafting work. Our position in this process was that we wanted to ensure that the city's planning stance was not undermined but was if possible reinforced through regional planning guidance. The Secretary of State's response (Government Offices for the North West and for Merseyside, 1995) came in the form of a consultation draft dated April 1995, which picked up much of the content of the advice submitted to it 12 months previously except perhaps in the field of transportation, whilst adding in some particular concerns of the government (notably the roll-forward of population figures and housing targets, and the desire to see a small number of 'flagship sites' identified in the region to enable it to compete more effectively for large-scale inward investment). Together, these two documents demonstrated the shift towards giving green issues a more central place in planning than had previously been the case (see Chapter 9 for a discussion of how emerging debates about sustainability began to impact on what was being done in Manchester), but they also arrived very late in the Manchester UDP process. In the event, this did not prove to be a problem in the sense that there was nothing in what was emerging here that couldn't be dealt with through the arrangements already in hand to keep the Manchester plan up-to-date postadoption, but the outcome could have been different and further time-consuming delays for the UDP process could have occurred if a revisit at a very late stage in the process prior to adoption had been necessary as a result of emerging regional planning guidance.

Conclusions

I have elsewhere (Kitchen, 1996a) argued that a minimalist test of the worth of a plan might be put by the following expression, which I have described as Kitchen's Law:

$$\frac{\text{Time over which plan is useful}}{\text{Time taken to prepare plan}} \geq 1$$

This is minimalist because it requires a plan to give back no more value than the effort put into getting it in the first place. None the less, many statutory plans do not pass this test. It is a daunting thought that on the basis of this expression, the Manchester UDP having taken five and a half years to get to the point of adoption by the middle of 1995 would have to survive intact until 2000/2001 to pass muster. Strategies for helping get near this kind of success measure include working to it for development control purposes immediately a deposit version is agreed, and developing relatively ongoing ways of keeping it up to date so that it is progressively adopted rather than standing or falling as an entity; both these, incidentally, probably require more sophisticated ways of formulating the above expression. None the less, these considerations do suggest that there are components of the British statutory plan-making process that could helpfully be re-examined to make it more cost-effective and to make Section 54A of the Planning Act (which broadly says that the development plan should have primacy in the development control process unless material considerations clearly indicate otherwise) more meaningful. These might include the following:

- The replacement of the existing panoply of legislation and guidance about the nature and purpose of statutory plans with a general enabling power to local planning authorities. After all, if the primary purpose of a plan is to help meet local needs, this should be the test of its worth in these terms rather than compliance with detailed national specifications. Do all plans really have to be broadly the same, using broadly the same approved language, or can we now accept that local diversity in these terms is not necessarily harmful, and that local planning authorities after being in existence for 50 years can broadly be trusted in these terms given appropriate safeguards? This is not an argument about the rights and wrongs of giving guidance about good practice in development plan-making (Department of the Environment, 1993a), because clearly there is a useful role to be played by the processes of distilling and communicating other people's experiences. There is a real debate to be had, however, about whether such activities should be used to inform a system characterised by diverse practice based upon local perceptions of need, or to seek to promote conformity through the creation of a universal 'DoE-speak'. A central authority often finds it very difficult to provide advice and then to leave it lying on the table for others to use as they see fit, but tends instead to try to use its own powers and influence to seek standardised behaviour. By such processes, useful advice can become a template, which can drive out the very creativity that was accessed to provide the advice in the first place.

- The abandonment by government of the notion that it has a gate-keeping role in relation to the process, and its replacement by a greater clarity about and a sole focus on substantive national and regional planning and related policies. This should be the real role of government, and not the sort of grannying about things which don't really matter all that much that I have described in relation to the Manchester UDP, which can all too easily become a substitute for a function that central government ought to perform but too often doesn't. Indeed, it could be argued that the Manchester UDP got exactly the opposite of what this model would demand; very limited substantive guidance about wider strategic matters and a lot more about essentially unimportant detail.
- The acceptance of the principle that at the end of the deposit period non-objected-to policies in plans, subject to adequate prior safeguards about public consultation in the run-up to this stage and the protection of the rights of *bona-fide* objectors, have the immediate force of law. This would allow the across-the-board benefits of plans to be experienced in parallel with their proper testing in contested areas, usually on a more limited number of matters, rather than for the former only to be secured after the latter as at present. It is worth remembering that most of the Manchester UDP was not objected to, and yet the benefits of statutory status for any of this can be held up for a long time by objections to unrelated policies in the plan because it has to be taken as a whole. It is arguable also that this present proposal would redress the imbalance between the rights of objectors to plan policies and the rights of beneficiaries from plan policies which has crept into the British system. At present, supporters of policies which are not objected to have no rights of appearance at the public inquiry into a deposit draft plan because it is an inquiry into objections, yet their interest in having policies adopted which they support or from which they may benefit is held in abeyance whilst objections to other matters are dealt with. It can be argued that this is an inequitable situation, which the proposal canvassed here would remove.
- The adoption by the inspectorate at inquiry of a much more inquisitorial role in relation to plans in place of the more passive quasi-judicial role that to date has been the norm. There are welcome signs that this is beginning to happen, but it could be taken much further.
- The acceptance in terms of the statutory force of the development plan that informal documentation that clearly sits within the provisions of a statutory

plan, intends to show in more detail in particular areas how the objectives of that plan can be achieved but is recognised as something which may have a relatively short shelf-life, and has been the subject of adequate consultation, should carry a status which acknowledges these relationships with the development plan. In other words, there is a case for examining in these defined circumstances the creation of a kind of quasi-statutory category of plan, as a mid-way area between the statutory or the non-statutory status that exists today. This idea would build on the concept of supplementary planning guidance with which there is already some accumulated experience.

Every one of these ideas would help to move towards a quicker-in and quicker-out development plan-making system. As with all proposals for change, each of these ideas could have counterarguments put up against it, and it may well be that on closer examination such arguments in some or even all cases ought to carry the day. But we need to look very seriously at this issue partly because the statutory process will become ever more protracted as a result of objections as the importance of this element of the process becomes more recognised, and partly because it must be a condemnation of statutory processes when local planning authorities seek to find ways of circumnavigating them because of their cumbersome and time-consuming nature. None of this, however, is to seek to take away people's rights in terms of the process or to devalue its worth as a framework for communication; rather, it is to try to sit these rights alongside the need to ensure that the process is reasonably effective in its own terms. As Healey (1994) argues, the contemporary challenge for the regulatory form of land-use planning systems is between forms of technicist management and forms of pluralist democratic management, and these changes may help to promote the latter without this necessarily being at the expense of enabling statutory plans to do a useful technical job.

5

Approaches to development control

Introduction and context

Development control is often regarded as a Cinderella element in the planning service, not least by planning staff wanting to do what they regard as more dramatic or exciting work less hemmed in by the law. Yet it is probably the part of the service that ordinary citizens are most likely to come into direct contact with, especially when it is remembered that the large-scale consultation exercises usually now associated with development plan-making typically generate very small percentage responses. To put this into perspective, we calculated towards the end of the 1980s that we sent out over 50,000 consultation letters in a single year as part of the development control process, which meant that on average every Manchester household would be consulted directly on a planning application between once every three and once every four years (since there are around 180,000 households in Manchester). Development control is also the part of the service which is most likely to be dealing with something that can have a very real effect (positive or negative) on the quality of someone's life in a locality in the near future, and it can as a consequence be an area within which passions are aroused and strongly worded views are expressed. This in turn performs a valuable role in both implementing and testing the content of the development plan. Finally, it is the part of the service which overwhelmingly dominates the department's reception facilities, because in a typical day the majority of customers at the reception counter are people who have come to look at submitted plans because they have received a consultation letter, developers or their agents coming in to carry out negotiations on an application, or people who wish to look at the Planning Register. These reasons mean that what has been said in Chapter 2 about the customers of the planning service has a particular resonance in relation to development control.

For an interesting general discussion about some of these issues in the field of development control, albeit one that is now nearly a decade old, see Harrison and Mordey (1987). Booth (1996) looks at some of these issues under the broad themes of certainty and discretion on an international comparative basis. A useful general text, with quite a strong legal component, which is a major element of development control practice, is Morgan and Nott (1995). An examination of some of the ethical dimensions of this and of the potential sources of conflict in development control is contained in Tewdwr-Jones (1995). In addition, the Audit Commission (1992) has published general guidance on the achievement of quality in development control, although it is possible to criticise this on the basis that it doesn't really take very far our general understanding of what the concept of quality actually means in relation to development control processes. The need for this was arguably subsequently thrown into clearer relief by the study undertaken by Audrey Lees for the Department of the Environment (1993b) of the north Cornwall case, where the development control process appeared to have gone seriously wrong. Like many other chief planning officers, I thought this report was of such significance that it warranted a report to my own committee because of the light it threw on member–officer relationships in development control.

Without repeating what has been said in Chapter 2, which has tried to focus on the perspectives of customers, this chapter looks at development control from the perspectives of the officers carrying it out and from the perspectives of elected members either as members of the Planning Committee or as ward councillors.

To put a broad scale against this, in the late 1980s and early 1990s Manchester typically received between just under 2,000 and approaching 2,500 planning applications each year, with the position in this range depending upon economic circumstances and upon fluctuations in what was or was not regarded at any one point in time as permitted development. The degree of variation between years could be quite marked for these reasons, and as Chapter 2 has already pointed out this could be quite a problem for the Planning Department when the forecast of the income that was received from planning application fees was an important component of the department's budget. The vast majority of these applications were small scale with only a very localised impact; typical of this would be a house extension application by a householder. A small number (perhaps in many years no more than 100 or so) were very large-scale applications with major and wide-ranging consequences to be understood and taken into account. As is explained later in this chapter, both the number of elected member meetings each year to take decisions on applications and the percentage that were dealt with through officer-delegated powers varied in the 1980s and 1990s, but to take typical figures of recent times if the council decided 2,000 applications per annum, with 65 per cent via delegated powers and with the Development Control Subcommittee meeting 12 times over the year to deal with the balance, this would produce an average committee agenda size of between 50 and 60 items.

In the light of these considerations, this chapter looks first at the views of the incoming Labour left-wing administration in 1984, since it was these views that in many ways shaped the development control process that exists today.

It then looks at subsequent changes in elected member decision-making arrangements and at how elected members relate to development control policy issues. The negotiation process and the closely related question of speed versus quality in development control are then examined, using some of the available comparative information on development control performance. Finally, a short section looks at the enforcement process, before some general conclusions are drawn.

The views of the incoming Labour administration in 1984

When the Labour left took control of the city as a result of the 1984 local elections, their view was that they wanted to open the development control system up so that it appeared to be (and was) more democratic. They also wanted to broaden out the consultation process on planning applications. Up to that point, reports on the Plans List (that part of the Planning Committee's agenda that deals with applications that are not regarded as being of such significance as to warrant full reports in their own right) had simply comprised blocks of text describing the application without a written recommendation. The recommendation to the committee through long custom and practice was actually given orally to it by its Chair, although it had in practice been agreed beforehand between the City Planning Officer and the Chair. There were two quite real difficulties (at least) with this:

- No one was really sure whose recommendation this actually was.
- No one really knew where they stood in relation to any application until the agreed recommendation was announced during the course of the committee meeting, by which time it was usually too late to do anything about it.

Whilst it was possible to argue that there were advantages inherent in both these difficulties, the former making the council's case at appeal easier at times because it didn't expose differences of view between members and officers and the latter because it restricted the scope for lobbying around individual applications, they epitomised to the incoming administration what was wrong with the process. Consequently, we quickly moved to a situation where Plans List reports adopted a much more formal structure of description–consultations–analysis–recommendations, which was widely regarded by all the parties to the process as a major improvement. This brought Plans List reports into line with main agenda reports on major applications which already did contain all this material including a full recommendation; it was one of the curiosities of the system that had grown up over time that written recommendations were presented in reports on major items but not on reports on more minor items.

Probably the most radical change proposed by the new administration, however, was the introduction of public speaking rights at committee; in other words, the right of interested parties to address the committee direct at its meeting about applications on the agenda. It has to be acknowledged that there was some senior officer scepticism about how this would actually work when the idea was first mooted, although there was also a belief in some quarters that it was an interesting idea well worth at least an experiment. The worries were essentially to do with how the process could be controlled, in terms of the total amount of time it might consume, the length of individual speeches and the scope for multiple (and largely repetitive) speeches on individual controversial items. The political will was to try to deal with these potential problems by agreeing and sticking to some simple rules; that individual speeches should be restricted in length to three minutes; and that where a large number of people turn up to speak on an individual item they appoint a spokesperson with speeches only being accepted after that on new points. At the time of writing, only two committee chairs (Arnold Spencer and Andrew Fender) have actually been responsible for administering this system over something like a 10-year period, and because they have both been committed to it but have equally seen the need to operate it with flexibility, tact and good humour, it really cannot be said that it has caused the major problem that some feared at the outset that it might. It has by now undoubtedly become a feature of the way Manchester operates its development control process at committee level.

Four dimensions of this are worthy of further consideration:

- To avoid people having to wait excessive periods of time for their items to come up (although of course this can still happen), the practice that has grown up is that all items on which members of the public wish to exercise their speaking rights are taken first. At committees where there are a significant number of items where this happens, typically this process can take up to 2–3 hours, and the consequence tends to be that the committee really whistles through the remaining items for reasons of fatigue if nothing else. Whilst it might be argued that this could mean that some items are not given proper consideration, elected members seem to have come to a pragmatic judgement as a result of experience that if no one wishes to speak on an item, and if the written representations are not too much at variance with the printed recommendations, and if the policy position appears to be clear-cut, then there probably isn't a real problem with a case.
- Secondly, whilst it could certainly be argued that there are practical limitations as to who is actually able to turn up and speak at a committee which begins early in the afternoon of a working day, this has not created a situation where the process is totally dominated by the middle classes able to get time off work because of the flexibility of their working arrangements which others may not have or by professionals addressing the committee on behalf of their clients. My experience on the whole was that the committee probably took most notice in any event of people who didn't fall into these categories, although these groups were often the people able to assemble the most organised presentations.
- Thirdly, we did a quick analysis after a year of operation of this scheme of what its effect on decision-making actually was, and what this showed was that in about 20 per cent of cases where speaking rights had been exercised

the committee's decision on that day was different from the printed recommendation. Usually, this took the form of a deferral (sometimes with a committee site visit) to allow us to seek to achieve some changes through negotiation that would help to ameliorate the problem, rather than an outright reversal of the recommendation which was relatively rare. One of the main benefits of this process was that it was very useful as a means of taking some of the heat out of potentially difficult situations, not only with major applications but also with neighbour disputes. To keep this in perspective, at the time that particular study was done about 55 per cent of applications were determined through delegated powers, so only about 45 per cent went to committee for determination; and of those about one-third at most were the subject of the exercise of speaking rights. It is this cohort to which the figure of 20 per cent with a different outcome refers, or about 3 per cent of the total number of applications at most.

- Fourthly, the emergence of these arrangements for securing public speaking rights dispensed over a relatively short period of time with arguments about whether the Planning Committee ought to handle all its business (including development control) at the same meeting or whether it needed to set up separate arrangements for handling development control. There is a very respectable intellectual argument to the effect that development control is an integral part of the planning process and should therefore be directly linked to other policy-making and implementation work in ways that are mutually reinforcing. This argument would lead to the view that committee arrangements should cover the full span of this work so that these connections can be made at elected-member level, and this was indeed the practice from 1984 for a while. What quickly happened, however, was that the exercise of speaking rights, and the decision that such items should be taken first at meetings so that people didn't have to wait to speak any longer than was unavoidable, combined to relegate all other matters to a much later place in the business of the committee than their placing on the agenda would suggest; and in practice these items often got short shrift because people were tired by then and simply wanted to get the meeting over. This would mean, for example, that a report about major policy matters would get less committee time than a development control item that had a neighbour dispute as an element of it. This raised some obvious questions about the committee's priorities. The consequence of this relatively quickly was the decision to establish a separate development control subcommittee, so that policy and implementation items could be dealt with by the main committee without the pressure of development control business determining how in practice they were handled.

The process of widening out consultation on applications took the form of a request to officers to exercise their judgement less conservatively when it came to deciding early in the life of an application which properties were likely to be affected by it. This has been a more or less continuous request ever since. The figure referred to above of more than 50,000 consultation letters being sent out in a single year was simply a snapshot which was not repeated in any systematic form. My estimate, from experience, was that this figure, which at the time that count was undertaken would have produced an average of 25 or so consultation letters per application but with wide variations around this figure because of the vastly differing circumstances of individual applications, was a considerable increase on what was being done in the early 1980s. It has since been extended further although probably by a lesser amount. However, we constantly came across people who thought they ought to have been consulted about an application but who lived outside the area defined for this purpose, and we also regularly came across people who claimed not to have received our consultation letter even though they lived well inside the defined area. This latter point is not really an indictment of the postal service in Manchester (the failure-to-deliver rate cannot surely be as high as this experience might suggest), but probably says much more about many people's reactions to receiving a letter from the council, which may be thrown away several weeks before a decision is made and some time also before it has become clear that the matter is the subject of some local controversy.

What all this indicates, however, is that the process of defining areas for consultation in individual cases is a judgemental one, and elected members on the whole accepted that this was so, once they could see that we were trying to carry out their general instruction on consultation. If they felt, by the time an issue came to be determined at committee, that we had seriously underconsulted, they would defer an application and ask us to consult more widely, but this did not happen very often. In all this, there were inevitably variations in practice arising from the fact that these judgements had to be made relatively early in the life of an application when its full significance might not be appreciated by staff, and also arising from the fact that these were the judgements of a large number of (sometimes relatively junior) individuals since they were usually made by the case officer for each application.

It is perhaps also important to record at least the key features of the development control process that the shift of political power in 1984 did *not* change. Four in particular have remained key features of the development control process in Manchester from the era before 1984:

- The department's commitment to a negotiating style (see below) was reinforced by the incoming administration because this produced both a better outcome and a better standard of service for our customers. Support for this stance has not wavered over this period, and the only issue has tended to be the council's policy parameters within which those negotiations have taken place.
- The need to acknowledge that representations on planning applications keep arriving right up to the point of decision. The fact that elected members want to know about all these in relation to applications in which they have taken a particular interest means that arrangements have to be made for reporting on this matter in as up to date a manner as is possible.

The arrangement that has remained in place for doing this is that a 'late representations list' is prepared no more than about 24 hours before the committee meets, which summarises and comments on the relevant material which has arrived since reports were finalised; and it should be remembered in this context that many such reports for all practical purposes are finalised nearly a fortnight before the committee meets. This late representations list goes into the party political groups which meet just before the committee, and is formally accepted as part of the committee's business under the Local Government (Access to Information) Act as an appendix to the main agenda.

- The department has for quite some time prepared a weekly list of applications received by ward, so that elected members and other interested parties could become aware quickly of what was happening in particular parts of the city and could organise themselves to get involved in the process of handling these applications if they wished. This has remained a staple feature of the department's information service, supplemented in recent times by similar lists prepared a few days before each Development Control Subcommittee meeting informing people of which applications were actually going for decision to that subcommittee.
- The need for a private preview arrangement between the officers and the chair and deputy chair of the committee has also been reinforced. Whilst there might have been some views to the effect that something like this didn't sit very comfortably with the stances on openness and democracy taken by the new administration, experience quickly showed that a preview arrangement (typically two days before the committee meeting) was useful both in helping those members to get up to date with issues and representations on difficult applications and in looking informally at the options that might be open to the committee in such cases. The value of the former point was simply that the printed papers would usually be something approaching ten days old by the time of the preview, and a lot of representations had often accrued over that period. The value of the latter point is, it is hoped, self-evident if the process of consideration of issues in private by the majority group immediately before the committee meets is to be as well informed as possible, especially since it can result in group decisions being taken in such meetings and being the subject of whips.

Changes in elected member decision-making arrangements

There were really two types of issues that occurred under this heading:

- Discussions about the frequency of meetings.
- Discussions about where responsibility for development control decision-making should actually sit.

The first of these can be explained very simply. In a system where some decisions are made by committee, the frequency of Committee meetings can be a significant element in statistics that record speed of decision-making (Audit Commission, 1992). By definition, a committee that meets once in a five-weekly cycle only provides half as many opportunities to bring an application to the point of decision as does a committee that meets twice in such a cycle, so up to a point a greater frequency of meetings ought to lead to an improvement in decision-making performance in terms of speed. On the other hand, more frequent meetings lead to shorter agendas for each meeting, which can create situations where the worth of the meeting is in question, especially since meetings cost money. In practice over the past 10 years this balance has resulted in shifts in Manchester between one and two meetings per cycle, with in recent times one development control meeting per cycle becoming the norm. This was at least partly as a result of corporate pressures to keep the number of meetings down. This has had two particular consequences. First, members have been more willing to accept items on the agenda which have not been finalised but which have been taken sufficiently far for the principles to be established and then for the final decision to be delegated to officers. The second is that the scope for increasing the share of applications delegated to officers has regularly been explored and has gradually been extended, so that the figure of 55 per cent of applications handled in delegation arrangements quoted above as being typical of the mid-1980s was by the mid-1990s, nearer to 65 per cent. Both these developments have meant that the question of the frequency of meetings has probably become less of an issue in Manchester than it once was, although it undoubtedly remains a factor in the development control performance figures (in terms of speed of decision-making) discussed below.

The history of where decision-making responsibility for development control actually sits within the council's committee structures has become complex since 1984, as compared with a situation in 1984 where all such decisions were taken by the Planning Committee either directly or by its Development Control Subcommittee once that was established. The main steps in this process of fragmentation can be set out as follows:

- In the mid-1980s, as part of the council's experiments with decentralisation, it was decided to delegate to the Wythenshawe Area Consultative Committee the right to be consulted on all planning applications in its area that were to go to the Planning Committee for decision and the right to approve (but not to refuse) those that were of purely local significance. Thus, applications that raised wider than local issues would come to the Planning Committee for decision accompanied by the views of the Wythenshawe Area Consultative Committee, and local applications that the consultative committee wanted to refuse would also come to the main committee sometimes with the recommendation of the consultative committee being in conflict with that of the City Planning Officer. A consultative committee for Wythenshawe had remained in place from an earlier attempt at establishing area arrangements of this kind in the city, perhaps because Wythenshawe saw itself in many ways as a free-standing township south of the River

Mersey with its own particular history and sense of identity; it was often referred to by members as 'the largest council estate in Europe'. At times, this created a situation of power without responsibility. The Wythenshawe Area Consultative Committee, which consisted mainly of all the elected members from the six wards in Wythenshawe, could play to the gallery in Wythenshawe and oppose applications that had attracted local objections without having to worry about whether that actually constituted a defensible basis for refusal by the council as local planning authority, because those decisions would fall to the Planning Committee under the scheme.

- The government established the Central Manchester Development Corporation immediately after the 1987 general election for an area covering approximately one-third of the city centre. It took as part of this process the power of development control authority, and an arrangement was negotiated whereby the City Planning Department became the Development Corporation's agent for development control purposes, although with no delegated powers of decision. This meant that, as City Planning Officer, I consulted the City Council about applications but actually made a recommendation to the Board of Central Manchester Development Corporation, which met in private without its papers being published as public documents and without any mechanism equivalent to the City Council's public speaking rights arrangement. The process of consulting the City Council did involve an element of delegation to me to express a view, essentially on the basis that a matter went to committee only if it would have been a committee item had the City Council maintained its development control responsibilities in this area. Where this process involved the formulation of elected member views, in the first instance this was done by the Planning Committee, but once the City Centre Subcommittee had been established (see below) it also got involved in this. As well as these decision-making arrangements, a consultation area was also defined around CMDC's area which enabled it to express views as a consultee about applications where the City Council was responsible for decision-making but which might have an impact on the Development Corporation's regeneration efforts.
- As part of the arrangements for pulling together decision-making in Hulme to complement the achievement of City Challenge resources for its regeneration starting in 1992/93, a Hulme Subcommittee was established as a subcommittee of the Policy and Resources Committee with delegated responsibility for most development control in its area. In addition, a consultation area around the fringes of its area was defined, essentially on the same basis as this had been done for CMDC as described above.
- Shortly after this, a City Centre Subcommittee of the Policy and Resources Committee was established to take an overview of actions in the city centre. It was not given delegated responsibility for development control decision-making but it became a consultee both for applications in the two-thirds of the city centre where the City Council retained development control responsibility and in the CMDC area.

- During 1995, as a result of a successful bid for Single Regeneration Budget resources initially for the regeneration of the Monsall area of north Manchester, broadly similar arrangements to those already extant for Hulme were established in respect of a North Manchester Regeneration Area. These delegated responsibility for most development control decisions to a new subcommittee of the Policy and Resources Committee, and along the lines of the CMDC and Hulme models a wider consultation area was also established so that this new subcommittee could advise the Development Control Subcommittee about matters potentially affecting the regeneration of north Manchester but outside its area.

This patchwork quilt of decision-making arrangements (see Figure 5.1) has arisen mainly because of the City Council's desire to integrate (some would say centralise) the way it handles major regeneration initiatives. There clearly are arguments in terms of the consistency and coherence of the regeneration process in favour of the view that the best way to achieve this is by locating relevant council responsibilities within specified individual committees, but as far as the development control process as a whole is concerned as compared with the very simple situation which existed in 1984 this has created a system which is

- complex;
- potentially confusing to interested parties, including elected members and staff;
- much more difficult to handle in terms of the achievement of consistency where this is needed;
- often more time-consuming because of the overlapping consultation arrangements between committees; and
- administratively complicated, in terms of routine matters such as deadlines, due dates for reports, and briefing and committee servicing arrangements.

The more recent components of all this have also raised concerns in some quarters about what this set of decisions implies about the role of the Development Control Subcommittee. It seems to imply that the subcommittee is to be marginalised in relation to major regeneration initiatives and left to deal with the bread-and-butter work elsewhere in the city. Whatever the rights and wrongs of this argument are (and these tend to depend on whether or not this is seen as a process of creeping centralisation), these recent arrangements have increased the potential for intercommittee conflict. This can leave the council appearing to speak with more than one voice, despite the fact that it is the council and not any individual committee or group of elected members which is the local planning authority.

Elected members and development control policy issues

As well as making decisions about applications which go to the Development Control Subcommittee for determination, one of its other main functions is to

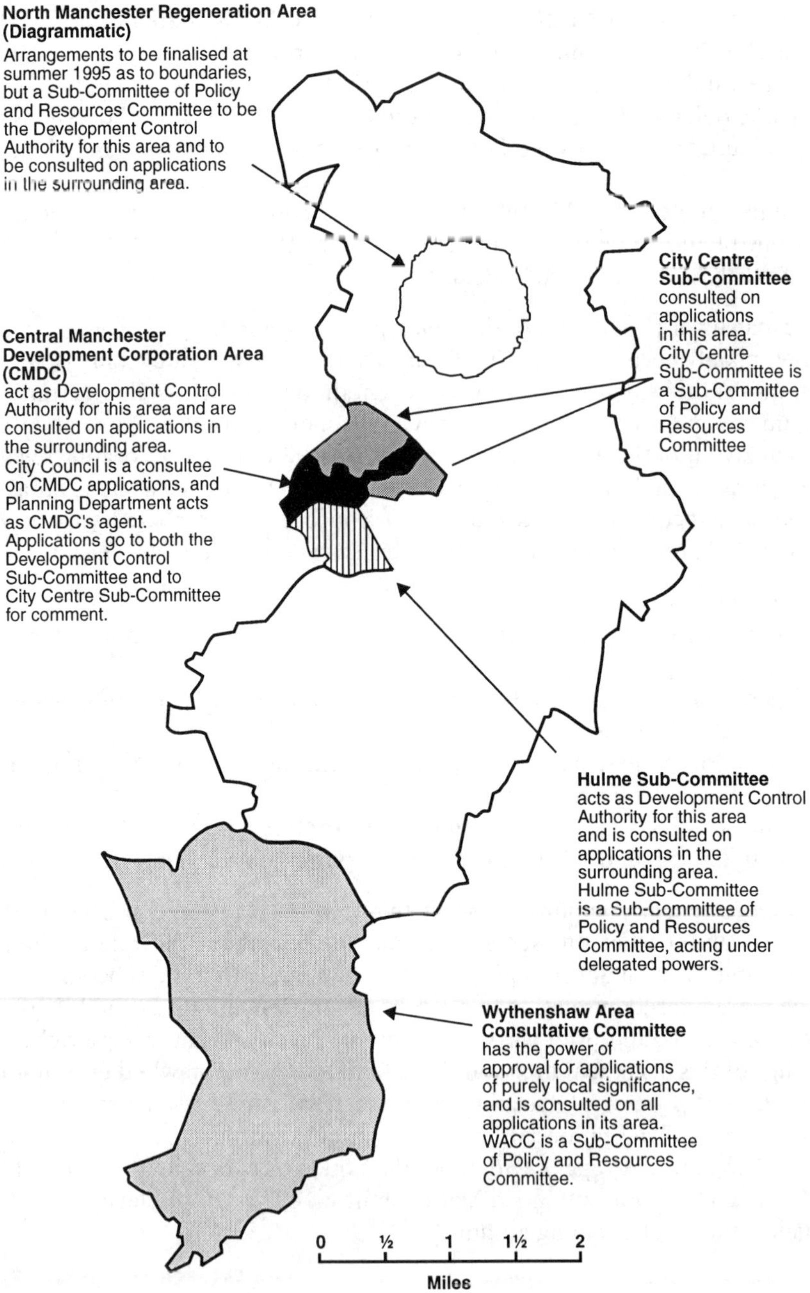
North Manchester Regeneration Area (Diagrammatic)
Arrangements to be finalised at summer 1995 as to boundaries, but a Sub-Committee of Policy and Resources Committee to be the Development Control Authority for this area and to be consulted on applications in the surrounding area.
City Centre Sub-Committee
consulted on applications in this area. City Centre Sub-Committee is a Sub-Committee of Policy and Resources Committee
Central Manchester Development Corporation Area (CMDC)
act as Development Control Authority for this area and are consulted on applications in the surrounding area. City Council is a consultee on CMDC applications, and Planning Department acts as CMDC's agent. Applications go to both the Development Control Sub-Committee and to City Centre Sub-Committee for comment.
Hulme Sub-Committee
acts as Development Control Authority for this area and is consulted on applications in the surrounding area. Hulme Sub-Committee is a Sub-Committee of Policy and Resources Committee, acting under delegated powers.
Wythenshaw Area Consultative Committee
has the power of approval for applications of purely local significance, and is consulted on all applications in its area. WACC is a Sub-Committee of Policy and Resources Committee.
0
½
1
1½
2
Miles

Figure 5.1 Development control arrangements at committee level at summer 1995

identify, as a result of these experiences, areas within which development control policies may need to be reviewed. There have been several of these over the years, many of which were subsequently incorporated into the work of the Manchester UDP described in Chapter 4. Examples drawn from the whole of this period include:

- Reviewing the problems of special needs housing when the amount of this which accumulates in a street or in an area is such as to suggest that the character of that street or that area may be changing as a consequence.
- Reviewing the extent to which the operation of the development control process appears to impact adversely on particular ethnic minority groups, and where this is demonstrably the case trying to ensure that policy guidance is targeted at these groups to seek to overcome the components of this that are caused by communication difficulties.
- Studying whether the continuing growth in the number of students living in parts of South Manchester was beginning to affect the character of such areas, and if so evaluating the extent to which the development control process could and should be doing anything about this.
- Reviewing what in practice can be achieved through the development control process to improve the accessibility of buildings to disabled people.
- Reviewing the operation of the development control process in respect of the provision of car parking spaces by developers, in the light of emerging debates about transport, the environment and sustainability.

This is just a selection of the sorts of issues of this type that have been tackled over the years, some rather more successfully than others. The general model for carrying out this sort of review work is:

- identification of problem;
- carrying out of appropriate in-house work;
- informal discussion with chair and deputy chair about what the results of this exercise might imply;
- carrying out appropriate consultations; and
- report to committee.

In the interim period whilst work of this nature is proceeding, there is inevitably a need to proceed with a degree of caution. Essentially the situation is that a problem has been identified but a solution has not yet been found, although one may be in the course of emerging. There would clearly be difficulties at appeal for the City Council if it took a decision in the light of what it thought might be emerging but was not able subsequently to back this up. But equally and understandably, the council would not want to make the matter worse. There really aren't any ground-rules that adequately cover this situation. Each case has to be taken on its merits. What all this does suggest, however, is that these kinds of policy review processes need to be concluded as quickly as possible to avoid any more applications being caught in the middle than is absolutely necessary. This was usually the principle on which

Table 5.1 Development control performance, Manchester and all metropolitan districts (%)

	Manchester	All metropolitan districts
Applications determined in 8 weeks	38	57
Applications approved	89	85
Applications that end up at appeal	3.5	3.9

we tried to organise such reviews. In many cases they will also need to be linked into the process of keeping under review the provisions of the development plan (described in the previous chapter), to ensure that significant policy changes are given the status needed to enable the Council to implement them with the maximum chances of success.

The negotiation process

Manchester has developed a negotiating style because it believes that this provides a better service to most parties to the process and because it also believes that this enables the end-product to be improved. The importance of this can readily be understood if you remember that most citizens probably judge the planning process, to the extent that they do this at all, primarily on the basis of the quality of buildings and spaces that get constructed. The effect of a negotiating style is probably most marked on the cluster of applications that at the time of submission might well have been refused if an immediate decision had had to be made. Its consequences are usually a slowing down of the speed of decision-making but a higher than average approvals rate and a lower than average appeals rate; and Manchester does indeed exhibit these characteristics.

For example, the 1993/94 Audit Commission profile of Manchester (Audit Commission, 1994), using 1992/93 CIPFA figures, shows Manchester in comparison with the average for all metropolitan districts (Table 5.1). In practice, in terms of changing the outcome, the willingness to negotiate impacts mainly on the pool of applications that are relatively marginal when submitted. Very broadly, about 80 per cent or so of all applications are likely to be approved anyway because they simply do not raise difficulties of a kind that would warrant a refusal; and this figure can be compared with the fact that around 65 per cent of applications in Manchester are dealt with by officers using delegated powers, almost all of which are approvals. Similarly, around 10 per cent of applications are likely to be refused anyway because in some significant ways they fall foul of current policy stances. The willingness of an authority to negotiate is thus primarily to be seen in how it handles the 10 per cent or so of applications that fall between these two extremes; this is shown schematically in Figure 5.2. In addition, of course, the negotiation process can achieve improvements to applications that are capable of being approved anyway, and these gains although not reflected in these figures can nevertheless be worth having in terms of the contribution that a building or a space makes to the local environment.

I have already commented in Chapter 2 about the general willingness of applicants to negotiate, and about the fact that this usually makes financial sense for them. I would also argue very strongly that this approach is more likely to produce a satisfied customer than an approach which puts less emphasis on negotiation, and the customer feedback figures from developers and agents also reported in Chapter 2 would appear to support this view. I certainly believe that as an approach it gives the city a better end-result, and it is

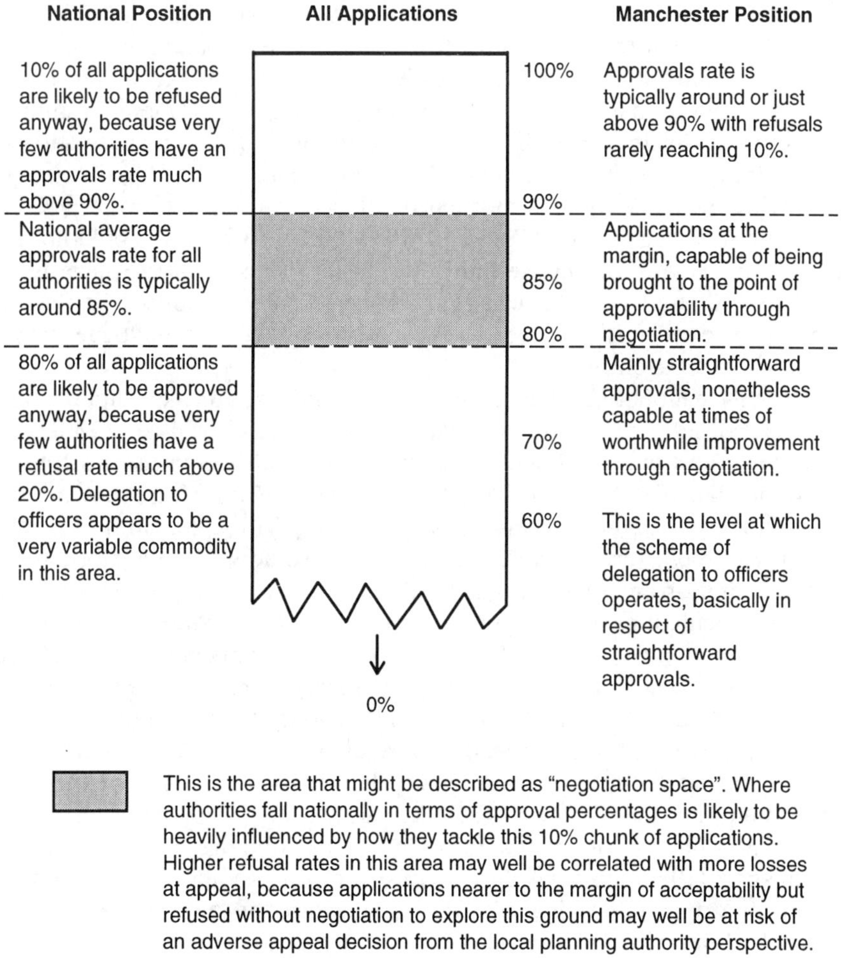

Figure 5.2 The development control pattern of decision making

Table 5.2 Development control comparisons, Manchester and all metropolitan districts

	Manchester	All metropolitan districts
Development control staff per 1,000 population	1.32	0.88
Applications received per 1,000 population	4.5	6.3
Average staff days spent per application	6.7	3.2
Average expenditure on the development control service per 1,000 population	£1,430	£1,160
Average cost per application	£316	£184

always worth remembering that we often have to live with the product of the development control process for a long time. At the same time, it needs to be acknowledged that there are limits to how far this process can go. Development control negotiations are not likely to turn a really bad design into a really good one, and in any event the broad guidance to local planning authorities from the Secretary of State for the Environment about these sorts of matters (see the annexe to *Planning Policy Guidance Note 1*: Department of the Environment, 1992) makes it clear that local planning authorities need to be careful in trying to tackle issues of this kind because they should not get involved in seeking to impose individual tastes. Thus, what we have to be content with, and what staff can gain a considerable amount of satisfaction from, is making worthwhile gains through a constructive dialogue with a developer or an agent.

Both this negotiating style and a heavy emphasis on consultation are, however, considerable consumers of staff resources, as well as of time; and these things cost money. Some figures from the Audit Commission's 1993/94 profile of Manchester (Audit Commission, 1994), again which compare the city with all metropolitan districts using 1992/93 CIPFA statistics, are quite revealing in these terms (Table 5.2). There are some well understood problems of definition and of consistency with the CIPFA figures that are the basis for these calculations, but taking them at face value they do indeed show a service in comparison with its family average which appears to be relatively slow, relatively heavy on staff resources and therefore relatively expensive. At the end of the day, these matters turn on decisions about the kind of service one is trying to provide; it would undoubtedly be possible to improve the above figures, possibly quite considerably, but this would be at the expense of negotiation and consultation. Where the best balance lies between service quality and value for money is at its heart a matter of judgement, as the Audit Commission (1992) acknowledges. Whilst the professional views of officers are an important determinant of this, and whilst the formal and informal feedback obtained from customers is also very significant, the final decision about what the service is trying to achieve ought properly to lie with elected members; and over the years the Manchester style has been consistently endorsed by the members of the City Council.

The negotiation process is going on all the time in relation to a high proportion of the development control workload, usually on a face-to-face basis between staff member and applicant or agent and often in response to comments that have been received through the consultation mechanisms that are standard practice with applications. A high proportion of this is amicable or at least non-confrontational, and only a small element relates to high-profile, very visible and very controversial issues. There are always judgements to be made about how far to go with activities of this kind, however; for example, is the effort that might have to be expended worth the gains that might be achieved? Is there a risk that the process might create a degree of confrontation that neither party really wants and that might actually make reaching a measured agreement on improvements harder rather than easier? Have we got the staff resources to pursue an intensive negotiation without causing too much delay to other customers? Does it really matter anyway in an individual case? These are genuinely matters of judgement in which the experience of staff is an important component, but the touchstones for these judgements tended to be whether or not there is a worthwhile gain to be had and/or whether or not there was a legitimate customer interest (for example, a present or possible future neighbour) to advance or protect. My view is that development control negotiation contributes considerably to the quality in the built environment that the planning process is able to help to achieve, and that more should be said publicly by the planning profession about its strengths rather than some of the defensive reactions provoked by the 'jobs locked up in filing cabinets' view of conservative government ministers during the 1980s.

Speed versus quality

Ideally, these two would not be counterposed. What everyone would ideally want from a development control service would be good decisions taken quickly. However, as I have tried to explain, policy attitudes about the nature of the service do affect its speed, and the resources that could help to reduce some of these tensions are not always available. We had a very dramatic illustration of this throughout the life of Central Manchester Development Corporation, which meant that through the agency agreement to provide a development control service for CMDC we could readily compare the approach in their area with the approach in the rest of the city centre and indeed with the rest of the city, since in effect these two approaches sat alongside each other and to a degree were operated by some of the same people. In looking at this comparison, it should be acknowledged that historically development control in the city centre tended to be a little faster than elsewhere, primarily because there was less local consultation with only a relatively small resident population. The effect of this was mainly in terms of the delegation system, since one of the triggers for an application becoming a matter for determination by elected members was the existence of substantive objections; and a higher proportion of delegated items tends to produce a faster throughput simply because there is no need in such cases to wait for the specific date of a committee meeting. The CMDC agency meant that development control in its area was relatively much better funded than elsewhere and worked to a streamlined decision-making procedure, because the board of CMDC placed considerable emphasis on taking development control decisions in eight weeks and organised (and sometimes reorganised) its meetings accordingly. The result was that typically CMDC's quarterly performance (although most of this was actually work done by Planning Department staff) would be at or in excess of 90 per cent of decisions taken in eight weeks, whereas for the rest of the city we would struggle to get the figure much above 40 per cent. Put more emotively, this tended to mean that on national published development control performance tables, CMDC was one of the national stars and Manchester City Council was one of the national laggards. Yet in terms of quality of output there simply was not this kind of visible difference between what were in many ways two arms of the same service, albeit one arm that was well resourced and another that was poorly resourced. This may in some ways be rather an extreme example, but it does show how much difference can be made by funding levels, by streamlined procedures targeted at helping to achieve an objective of this nature and by more frequent decision-making meetings.

This is not to say that in the rest of the city it was not possible to improve performance. Speed of decision-making can properly be regarded as a component of quality, and our customer feedback showed that many customers thought we were too slow. As a consequence, and partly in response to national pressure from the Department of the Environment from the early 1990s, we looked at what we could do to move towards achieving 50 per cent of decisions within 8 weeks without compromising the essence of the service on offer. The main areas we focused on were trying to take a clearer view about when negotiations ought to stop because there was little worth while to be gained from their continuation, and trying to deal more expeditiously with those items of delegated decision-making which were straightforward. The effect of this was that performance did indeed begin to inch towards the 50 per cent target as compared with a 1992/93 level of 38 per cent reported above, without in my judgement prejudicing what we were trying to achieve. What we were not able to tackle, however, was the basic staff resource position, because of the overall budgetary position described in Chapter 1, and thus this attempt to improve the development control performance figures was essentially a management exercise within existing resources. It is perhaps worth noting in this context that in the mid/late 1980s, with higher staffing levels, slightly higher application levels, but a lower percentage of applications determined through delegated powers, the department was able to achieve figures of around 60 per cent in its best years.

To my mind, all this has taken on a significance in recent years that it does not warrant, as speed of development control decision-making has become the published performance indicator for planning authorities. I do not advocate inefficiency, but the issue of speed needs to be kept in perspective. It is easy to take a bad decision quickly and to repent at leisure, but it is doubtful whether most of our customers would regard this as a good standard of service. More emotively, the buildings of today will not be judged in years to come by whether or not the applications that led to their appearance were determined in 8 weeks rather than in 10; indeed, everyone will have forgotten this long before the time for such judgements has arrived. I would certainly prefer to contribute constructively to the achievement of some good-quality buildings than to worry excessively about speed; in future plaques may be put on some of today's buildings to acknowledge the fact that they have turned out to be the listed buildings of tomorrow, but it is not likely that a plaque will be put on a building to commemorate the fact that the planning application for it was determined in less than 8 weeks. Despite the fact that this is manifestly not a performance measure for the whole of the planning service, in terms of the statutory performance indicators on which local authorities have to publish information this remains at the time of writing the only criterion that is set for the planning service. In addition to all the other things said above about this indicator, it is clear that it is not a measure of quality, except in so far as it can be argued that speed of decision-making is a component of a quality service. I prefer to believe that it exists as a sole public criterion because no one else has yet devised a more appropriate one. If true this is a classic illustration of how the measurable can get elevated in importance just because it is measurable. One of the challenges posed by John Gummer's quality initiative (Department of the Environment, 1994b) must be to develop ways of expressing the achievement of quality development that can sit alongside an indicator of speed, to begin to express more effectively what the outputs and outcomes of the planning service actually are.

Enforcement

A particular segment of the development control process that can at times attract considerable media attention (often along the lines that the powerful local planning authority is giving a hard time to an individual who happens to have fallen foul of it) is the enforcement system. In Manchester, in a typical year we would get between 400 and 450 cases that might potentially raise enforcement issues drawn to the department's attention, often as a result of approaches from third parties and sometimes as a result of what our staff had themselves seen when out on site. Every such case was recorded and investigated. In the vast majority, there was either no breach of planning control or the issue was so minor that it was a relatively straightforward matter to deal with. In only a tiny minority were the issues both so serious and so incapable of being dealt with by negotiations that they ended up as court cases. In a typical year, from the potential caseload described above, only perhaps 10 or 12 would end up in court, or no more than about 2–3 per cent of the total. This result too reflected the Department's negotiating style; our attempt was to find a satisfactory resolution where we could, rather than to end up in court as quickly and as often as possible.

Wholly typical of this sort of pattern would be the individual or organisation that had breached planning control without being aware of it, and nearly all such cases were capable of being regularised usually with the very willing compliance of the individual or organisation concerned. Some of the breaches were knowingly undertaken, however. A typical example here would be a hot-food take-away operating late at night beyond the opening hours specified in its consent, where often the threat of legal action would suffice but where if it didn't the matter would need to be followed through. The hardest cases, and as I have already indicated these constituted a tiny minority of the total, involved people or businesses who were perfectly clear about what they were doing and who had every intention of using the delay capability inherent in legal processes to continue doing it for as long as possible. This was very frustrating for everyone concerned because it could drag on for years, and it was particularly difficult in terms of the department's relationships with its customers living or working in the area immediately surrounding the offending use, because they typically tended to think that the Planning Department had the power to close down the offending operation more or less immediately and were often not very receptive to explanations as to why this was not so. The law in these terms has been improved in recent years, but like many other practitioners I don't believe that we have yet got to the point where a proper balance has been established between the various interests in these sorts of circumstances. From the perspective of business, however, a national study sponsored by the Department of Trade and Industry (Interdepartmental Review Team, 1994) concluded that there were no strong arguments for fundamental changes to the system.

By its very nature, some elements of the enforcement service require not only a quick response but also observation work outside normal hours, and both of these things are demanding of staff resources. Because of this, and because if anything the number of potential enforcement cases grows with the increase in the number of community groups acting in effect as the eyes and ears of the planning service, we managed by moving money around within our budget to increase the number of enforcement staff from three to four during 1994; and the increase was actually more dramatic than this, because the number of people in post had dropped to two as a result of an unfilled vacancy. This is none the less a very small number of people to be handling the scale of caseload described above, and thus the work of the enforcement officers is very dependent both on good teamwork with other members of planning staff and a co-operative relationship with third parties and particularly with those who make a complaint in the first place. In this context, enforcement can be seen as part of the negotiating and consultative style of the department, and whilst this might appear to be a contradiction in terms it is worth remembering that if 10 out of 450 cases end up in court 440 do not but are resolved by other means.

Conclusions

Typically, the general elements of the development control process which cause elected members on committees that determine planning applications to feel that they are tightly constrained are the relatively narrow basis that exists in law for the determination of submissions and the slowness of the enforcement process in putting to right breaches of that law. On the first of these matters, it is clear that applications have to be decided on planning grounds rather than on any other basis; for example, the effect on property values in the locality is not a planning consideration, but it can be a cause of considerable local concern; and the characteristics of the occupants of a property as individuals are not generally a planning consideration, although again this can be a cause of considerable local concern. The move in recent times to reassert the importance of the development plan through Section 54A of the Planning Act can be both a help and a hindrance in these terms. It can be a help because a clear and up-to-date development plan sets down a framework of appropriate policies for decision-making, and whilst it is not a question simply of reading off the answer from the development plan (because each case must be considered individually) there is no doubt that it gives some strong leads in very many cases. It can be a hindrance because in a sense it can reduce flexibility. If the starting point is the provisions in the development plan, a council that chooses to depart from this position in making a development control decision is clearly at risk at appeal of having those very provisions argued against it, and planning officers need to advise their committees about such matters and about the attendant risks such as costs awards against the council even when they know that the stance the committee wants to take is politically popular and hence that officer advice will be politically unpopular. As far as the second area of typical elected member dissatisfaction with the development control process is concerned, the slowness of the enforcement system when it involves all its stages does to my mind place the development control system in some disrepute with some of its customers; it seems unfair that people who are determined to do so can get away with breaches of planning control for protracted periods of time. There is, therefore, a case for taking the reforms of recent years further, and actually looking at more summary forms of enforcement along the 'fixed penalty' lines found in other branches of law. It needs to be remembered, however, that only a tiny minority of enforcement cases go through all the potential stages, and the vast majority are cleared up in effect as rather a specialised arm of the negotiation process. Further reform should not have the effect of damaging this kind of exercise of discretion.

The development control process is probably the part of the planning service that is most involved in a continuous interaction with its customers. My experience is that most of these want to see good-quality development in the right places as the outcome, and want the process to inform them adequately, to consult them properly, to listen to their views carefully and to try to achieve an outcome which responds constructively to what has been said. Some of this is obviously mutually exclusive; implacable opposition to a development proposal by local residents is not likely to lead to a meeting of minds between them and the developer. But whilst some of the development control process is as black and as white as this, much of it can be found in the various shades of grey in between, and it is in this area that the processes of consultation and negotiation can achieve most. Much of this is not always easy to see; an undistinguished building will be judged as this by the people who have to live with it daily, and they are not likely to moderate this judgement by acknowledging how much better it is than the original submission. None the less, the accumulative effect of all of this on our environment can be quite considerable. It is hoped that this chapter and Chapter 2 have given an idea of these complex and dynamic processes, which certainly don't deserve the tag of 'Cinderella' that I labelled them with at the start of this chapter.

The material on customer feedback both from the Manchester survey and from the national project summarised in Chapter 2 appears to show that the development control process is quite widely understood and quite well accepted in terms of the functions it performs, and often also is seen as providing a good level of service to its various customers. Judgements about this need to be made carefully, however, because satisfaction with the operation of the process and satisfaction with the decision are clearly analytically separate ideas which often seem in practice to get mixed up in people's minds when they are talking about development control. This is an important distinction to make in terms of a service which produces an outcome (a decision) in relation to which there are often winners and losers. Alongside this very high level of visibility in terms of outcomes can be a very low level of visibility in terms of processes. An important characteristic of the service is that an emphasis on negotiation, as is to be found in Manchester and in many other authorities, often leads to significant improvements in terms of a built-form outcome; but people do not see the degrees of improvement that the efforts of the service have helped to achieve because they do not see what would have been built without those efforts. These gains are difficult to measure at any rate in any overall sense of assessing the performance of the service. One of the greatest satisfactions for staff can be in contributing to this part of a real improvement to a submitted scheme; but many of the customers of the planning service will not be aware of any of this and will simply respond to the final outcome. This in itself is a powerful argument for effective development control.

PART III

THE MAIN ARENAS

6

Manchester's economy

Introduction

My perception is that contemporary Manchester has six major economic strengths, and while it doesn't use precisely this language this is also essentially what the City Council says (Manchester City Council, 1994a). This chapter focuses on these six:

- The range of higher-order services in the city centre.
- The Higher Education Precinct.
- The city's transportation nodality.
- The range of high-technology activities in the city.
- The city's cultural and sporting life.
- The city's people.

The thinking behind this approach is that if cities are to nurture their economic base, they need to understand what their assets are, and then to seek to make them as strong as possible; and the planning service must be an integral part of this process. This kind of inventory is radically different from that which would have been compiled for Manchester a century ago, when the city's economic life was still trading on its role as the 'shock city' of the Industrial Revolution (Briggs, 1982, pp. 88–138). It would not be unreasonable to speculate that the inventory that might be compiled in a hundred years' time might also be radically different from today's. The point is that cities are in the business of economic change. They came into being in the first place on the back of economic changes, and they will continue to have a role only in so far as they have things to offer that are relevant to contemporary life.

The steam railway locomotives that Bayer-Peacock used to produce in Gorton in East Manchester, and that *aficionados* can still see operating on some of the far-flung railways of the British Commonwealth, may well have been things of beauty and wonder; but no amount of nostalgia for a return to this sort of world will create a market for products which are now obsolete. It is no surprise that the company in its original form no longer exists in Gorton.

We may accept that local government and planning activities cannot stop these kinds of structural changes from occurring. However, this does not mean that other things cannot be done to help cities to play to their strengths. This is the main emphasis of this chapter.

This optimistic view of the world presupposes that cities are capable of continuing to survive and perhaps even to flourish. This almost certainly won't be equally true of all cities but there is no reason to believe that the sorts of threats that cities face today need necessarily be terminal – if they have an asset base that is strong and relevant to the world in which they must live. Nor need city councils be passive in the face of what might be seen as inevitable decline; this chapter shows that constructive response is possible. None the less, the threats are both real and powerful. Examples include

- the structural decline of basic industries;
- the decentralisation of economic activity;
- congestion in all its forms; and
- developments in information technology which reduce the historic emphasis on the collective workplace.

Separately and together, these are powerful forces which will continue to impact on cities. But cities have to be able to compensate wholly or partly for the adversities they face. Economic activity in all its forms is the *raison d'être* of our cities; what has changed for Manchester is not this proposition, but the nature of the economic activity itself. The nature and the scale of the change from an economy based upon primary and manufacturing activity to one increasingly based on services, is dramatically illustrated in some of the work originally carried out for the first Greater Manchester Structure Plan, which presented a snapshot of what had been happening over a mere 13-year period (and thus was a fragment of a long-term process) in an area known as the Manchester Local Labour Market Area. This was much larger than the administrative City of Manchester (Greater Manchester County Council, 1975). The problems of producing completely comparable figures for jobs in a location over a period of time apply to these figures as well, but the broad picture they present is not significantly distorted by these difficulties (Table 6.1). Over this 13-year period, primary industry (mainly in coal mining) virtually disappeared, nearly 150,000 manufacturing jobs were lost, and the service sector, although growing by about 80,000 jobs, was able to make up just under half of the total number of jobs lost. Although directly comparable figures are not available, the position in the administrative City of Manchester, with around one third of this total number of jobs, would have been a more extreme version of this situation. As the core city, the services sector would have been stronger and the primary and manufacturing sectors weaker, but broadly the same trends would apply.

A very visible manifestation of these sorts of changes during this period was the fate of the Manchester Ship Canal, itself a symbol of the city's sense of its own commercial importance during the last part of the 19th century. Contemporary

Table 6.1 Jobs in the Manchester Local Labour Market Area, 1959/72

	1959	Share of total (%)	1972	Share of total (%)	1972 in comparison with 1959 (1959 = 100)
Primary	24,421	2.4	5,376	0.6	22
Manufacturing	533,648	52.2	387,647	41.4	73
Services	464,192	45.4	544,112	58.0	117
Total	1,022,261	—	937,135	—	92

photographs from the *Manchester Evening News* show that the canal was full of cargo ships as late as 1957, but by the late 1970s/early 1980s it had virtually no traffic and therefore had largely ceased to function as the channel serving a major inland port (*Manchester Evening News*, 1993). It is perhaps hardly surprising that changes as dramatic as this led the council to take a pessimistic view of the city's economic progress for much of this period. It is difficult to see how others could be expected to have confidence in the city, when the City Council itself did not always appear to behave in this manner. Looking back, my perspective is that it was not really until the 'City Centre Manchester' campaign of Christmas 1982 (a joint effort between the City Council and the Chamber of Trade in which the Planning Department played a major part) demonstrated that it was possible to fight back against adverse trends, that the City Council started to believe that continuing decline was not inevitable. But I am not sure that it was obvious at the time that this was to prove a watershed. Since then, however, the City Council has taken an increasingly proactive view of these things, and there appears to be independent research evidence that Manchester in recent times has fared rather better than many of its comparator (and competitor) cities in England (Robson, Bradford and Tye, 1995, pp. 123–35).

This chapter, in focusing on the city's six major strengths and on what has been done to enhance them, illustrates the processes at work, and the contribution that the planning service can make to them. Quite a lot of the material presented here is descriptive, but this is because it is important to understand what drives the economy of a city if planning for it is to be effective.

The range of higher-order services in the city centre

The concept of the regional centre goes beyond what would conventionally be defined as Manchester city centre and also encompasses the Higher Education Precinct, parts of inner Salford including Salford Quays and Salford University, and the area of Trafford Park Industrial Estate known as Trafford Wharfside (see Figure 6.1). Because this concept covered parts of the areas administered as local planning authorities by three district councils (Manchester, Salford and Trafford) and by two development corporations (Central Manchester and Trafford Park), it would have been impractical to focus resources on seeking to achieve a shared, detailed planning framework across

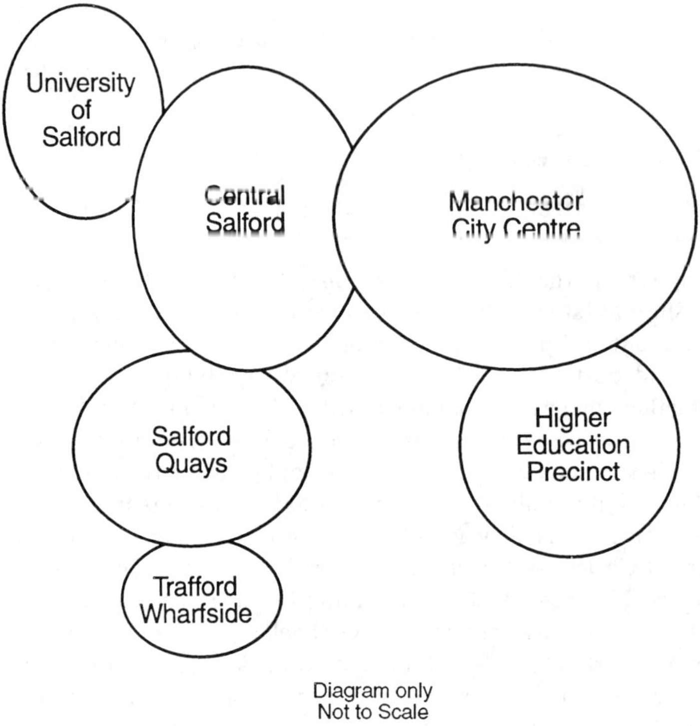

Figure 6.1 The concept of the regional centre

the whole area. None the less, it was important, in planning terms, not to allow thinking to become constrained by these administrative complexities. For example, the boundary between Manchester and Salford may go down the middle of the River Irwell, but a shared view about the river and its banks as an entity needed to be reached between the two authorities, as indeed it was. A consequence of this complex situation was that, as an early part of the process of working on the Manchester UDP, an outline strategy for this wider area was assembled in the City Planning Department and was sent to colleagues in the other four organisations to try to ensure that our basic thinking about this wider area was both accurate and shared. This was subsequently mirrored in the approach that emerged to City Pride (see Chapter 7), which again demonstrated that the strength of the regional centre as a concept was not limited by administrative boundaries.

In practice, what this concept means is that, within this area, the major regional functions of government, the major banking and financial institutions and professional services, a major concentration of retailing activity, concentrations of arts, leisure and entertainment facilities, and the clustering of higher educational institutions are all to be found. Much of this is dealt with in more detail later in this chapter. Taken together, however, what it all means is that about 100,000 of the 300,000 jobs to be found today in the City of Manchester are in the city centre, and if the area is widened to include the areas described above as the regional centre, the figure would be much larger than this. This relatively small area is the largest single concentration of jobs in northern England, and the critical mass that this provides is vital to the city's economic future. Because of this, planning in the city centre is a high-profile activity, and one which is distinct in several ways from that which is done elsewhere in the city. Thus, the work of the city-centre area team tended to be somewhat different in character from the work of the Planning Department's other area teams, with a clear strategic context which recognises the economic importance of the city centre provided by the City Centre Local Plan (Manchester City Council, 1984) and an emphasis on fine-grained work in each of its localities often with a strong design input, since most of the city centre consists of conservation areas and contains many listed buildings. Figure 6.2 subdivides the city centre into 10 subareas, each of which has a distinctive character, and in each of which the planning approach needs to be adapted to the requirements of this particular character.

A significant proportion of the jobs available in the city centre are office jobs. The trends affecting office locations in the Manchester area in recent years have been summarised by Haywood (1996). He shows how the booms and slumps of the property development market affected not only completions over the period 1985/93, but also their geographical location, and he relates these results to other considerations such as car parking provision and transport policy, and the emergence of the B1 use class in 1987, which combined the previously separate light industry and offices use classes into one. Because of this, the need for planning permission for changes of use from light industrial to offices was removed. This made it harder for the planning service to maintain office location policies. The decentralisation of office activity from the city centre had been around long before the creation of the B1 use class, of course, and it is possible to speculate endlessly about the future locational flexibility that developments in information technology may produce. But what the city centre continues to have going for it as an office location are the benefits of very substantial clustering, and the range of ancillary and support services that it offers. And while the planning service will always struggle to influence the internal logic of the individual locational decisions of companies and organisations, through its efforts in relation to planning the city centre as a whole it can influence the quality and the perceptions of what is on offer as a collective good. This is just one reason why planning in the city centre has an important role to play in the continuous effort to secure the economic base of the city.

Many of the jobs provided in the regional centre reflect its role as the major retail centre for the conurbation and for a large part of the region. This role had been significantly affected by the very lengthy process of assembling land for, and then constructing, the Arndale indoor shopping centre. This process

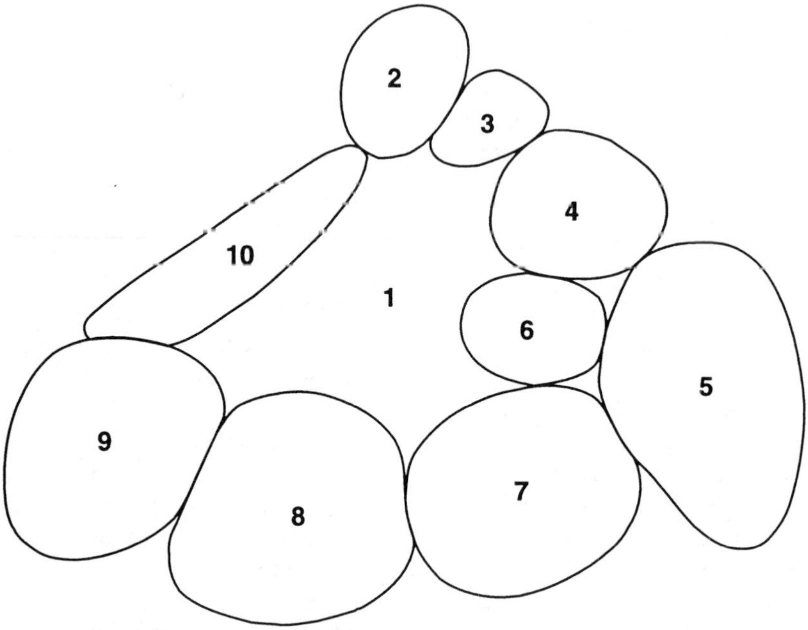

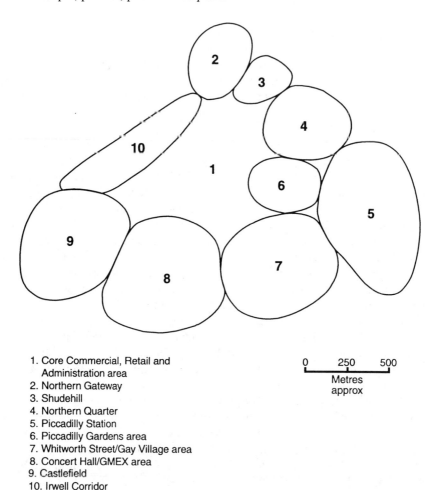

Figure 6.2 Manchester city centre – sub areas

lasted from the late 1950s to beyond the middle of the 1970s. During this period, many other surrounding shopping centres redeveloped more quickly, and took trade that had historically come into Manchester by bus and by train. This was part of a growing dependence on the private car for much retail access. The 1980s was a long period of struggle to consolidate, helped by extensive advertising campaigns around the Christmas period, and by the pedestrianisation of Market Street and St Ann's Square in the 1980s and their redesign in the early/mid 1990s. The 1990s have seen two further threats. The first of these has been the long-drawn-out saga of whether a major new regional shopping centre should be allowed at Dumplington, in the western part of the main area of Trafford Park Development Corporation and only a few miles to the west of the city centre. After a protracted fight at planning

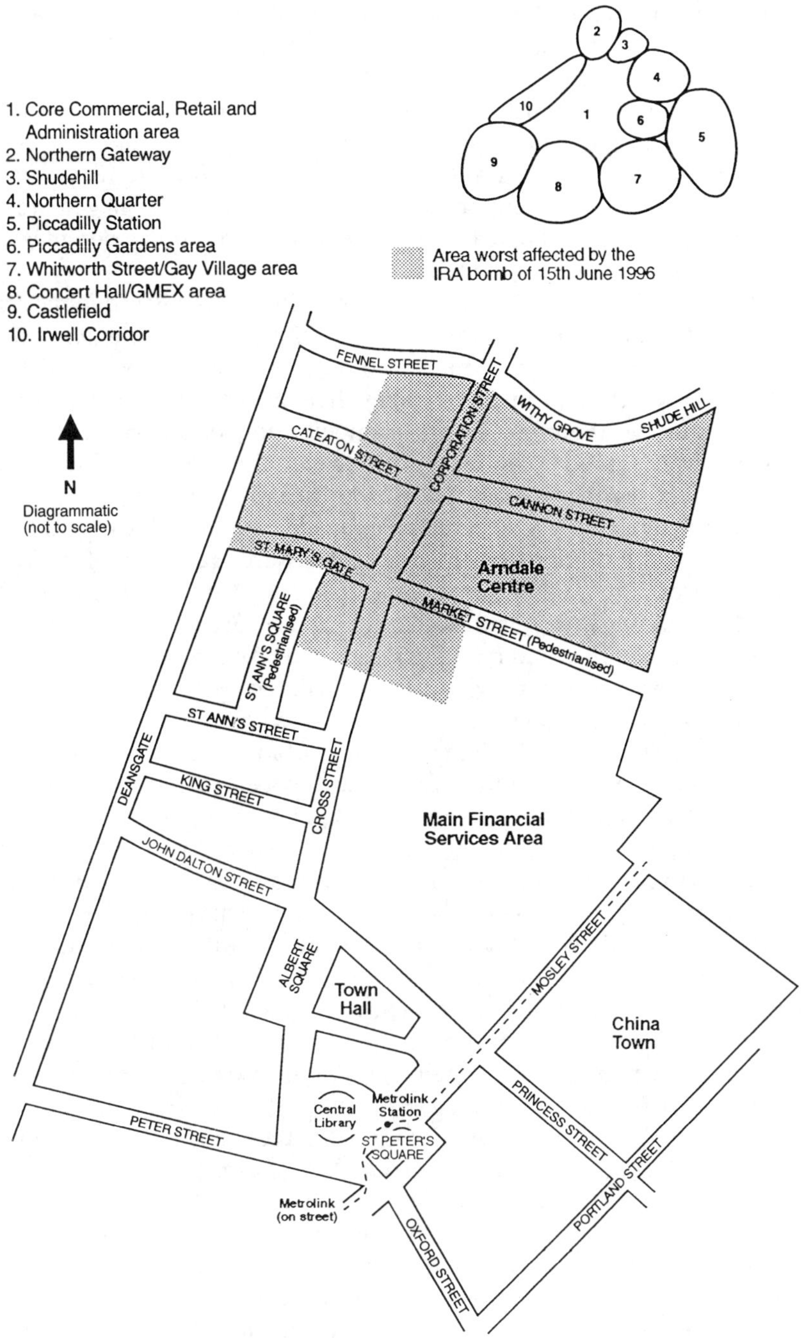

Figure 6.3 The administrative and commercial core of Manchester City Centre

inquiries and then through the courts as far as the House of Lords, the centre was allowed; and it clearly represents a challenge to the city centre's retail function. The second threat was the IRA bomb of June 1996, which destroyed or severely damaged much of the city centre's prime retail floorspace (see Figure 6.3). The combination of these two threats will undoubtedly affect the city's retail performance for a period of a few years. But the IRA bomb also presents the city with an opportunity to redevelop badly damaged areas to a better standard. This is being tackled though an international design competition, which is underway at the time of writing.

In thinking about the role of the city centre in generating economic activity, however, it is important to stress that it is not just a working environment where people commute in, work the conventional office day, and then commute back home again in the evening. It is also a place where visiting of various kinds is important to the economy, and increasingly, a place for people to live. This adds considerably to the vitality of the city centre as a mix of uses operating increasingly on a 24-hour basis. There is a great deal of economic activity that is dependent upon this spread which would not be there or would not flourish if the city centre were merely an office location. These matters are dealt with at some length in Chapter 9, but it may be helpful at this stage to quantify the importance of visiting and of living to complement the office and retailing jobs in the city centre.

As far as visitor-based activity is concerned, figures from the Greater Manchester Visitor and Convention Bureau (Greater Manchester Visitor and Convention Bureau, 1995) show how important tourism has become particularly to the economy of the city centre. City-centre hotels saw their weekday room occupancy rates go up from 67.2 per cent in 1991 to 72.8 per cent in 1994, and their weekend rates increase from 46.0 per cent to 51.6 per cent over the same period. Traditionally, weekday room occupancy has been supported by business life in the city centre and the weekends have been relatively slack, but these figures show that both have been buoyant in recent years.

A large part of this is from UK residents rather than overseas residents. Although the unit spend of overseas visitors tends to be higher, 61.5 per cent of Greater Manchester's tourism spend in 1993 was from UK residents. Of this overall tourism spend, the main parts were accommodation (35 per cent), eating out (24 per cent), shopping (19 per cent), and travel in the UK (14 per cent), showing in practice how widespread is the impact of tourism expenditure on the economy. Of the total number of overseas visitors staying in Greater Manchester, approximately three-quarters stay in the City of Manchester. By 1993, according to the British Tourist Authority, Manchester had become the sixth most popular location for overseas visitors staying for at least one night, ahead of more traditionally recognised 'tourist' cities such as York, Bath and Cambridge. When day trips to the city's various attractions were added in, these figures show a large and flourishing industry. Tourism to Manchester is not a music hall joke, as it was originally perceived to be when the City Council first started taking it seriously as a policy area in the late

1970s/early 1980s. It is a major part of the city's economy with a very particular city-centre dimension to it, because of the concentration there both of attractions of all kinds and of hotel bed-spaces. One of the consequences of this over the past decade or so has been an increase in hotel bed-spaces and a significant programme of upgrading many of the city centre's older hotels. This phenomenon of recent tourism growth is not, of course, unique to Manchester but appears to be both a common target and a common achievement for many large cities (Law, 1993).

Considering the city centre as a place to live, although there were a small number of earlier initiatives, this only really became a significant element of city-centre life in the second half of the 1980s. Until that time, the Planning Department had been arguing for it in many quarters and it had been a feature of the City Centre Local Plan, but it had been really difficult actually persuading developers to do anything very significant about it. To keep the policy alive, the department had often fought hard to secure a very small number of housing units in commercial schemes as part of the negotiating process around planning applications, and had had to settle for this as the best that could apparently be achieved. It was not able to claim significant successes until the potential for many of the cotton warehouses in the city centre for housing conversion, and the scope for new housing alongside the city centre's waterways, was recognised by private developers and by housing associations. They in turn needed to perceive that a market could be created to occupy such accommodation. Grant regimes that assisted with this process were an important part of this revitalisation, and it was one that the newly arrived Central Manchester Development Corporation was well placed to exploit. A residential population of little over 1,000 in 1991 had become about 3,000 by early 1995, and was likely to reach 10,000 by the millennium or just afterwards (Central Manchester Development Corporation, 1995).

The importance of this for the city centre was not only that it added to the range of activity at the evenings and at weekends, but also that a resident population was beginning to take a view about its immediate environment which was clearly different from the sorts of views that visitors express. People who choose to live in the city centre do so for a variety of reasons to do with their own life-styles, and have to accept in so doing that it is a city-centre environment with the range of uses and types of activities one would expect to find in a city centre and is not a suburban environment. A resident population on the scale expected early in the next century will have an effect on policy in relation to the city centre that has not been seen to date, although such perspectives would be familiar in many of the European cities that do have a tradition of city-centre living, which Manchester aspires to emulate.

The Higher Education Precinct

In many ways the Higher Education Precinct (which geographically forms a southern extension of the city centre) can be thought of as if it were part of the city centre. Certainly in formal development planning terms, this relationship is reflected in the Manchester UDP. The students and the staff attracted to the three universities (Manchester University, the University of Manchester Institute of Science and Technology, and Manchester Metropolitan University) and to the associated hospitals and other further and higher education facilities in the area, are a major part of the growth in city-centre housing referred to above, and there is a symbiotic relationship between the universities and many city-centre facilities. At the same time, in economic development terms it is helpful to think of the precinct separately, because there are significant differences between its character and that of the city centre as a whole.

During term-time, the day-time population of the precinct is approximately 50,000 people. Manchester is fond of saying that this makes it the largest single higher education campus in western Europe, and whether or not this is true it does provide a sense of the relative importance of the precinct as an asset to the city. This works at two levels at least. The first is that this is a very significant concentrated spending power in its own right. In addition, the spin-off effects into the local economy of the availability of such a concentration of academic expertise are already considerable. Robson *et al.* (1995) have shown that the four Greater Manchester universities (the three in Manchester plus Salford University) were responsible in the 1992/93 academic year for about £450 millions of expenditure per annum in the economy of the area, plus their capital expenditure programmes. But what is of even greater significance is the potential for the future contribution to the local economy directly and indirectly which could be made by the major institutions in the precinct as they continue to grow. It is probably a truism that the full extent of the scope for a university to contribute to local economic life has yet to be understood in Britain, not least by our universities themselves; but equally, local authorities need to understand that boosting the local economy is not the only business of a university.

Whatever the precise balance between these arguments, it is the potential inherent in the roles of the city's three universities as key economic players that has led, over the years, to the emergence of constructive working relationships with the City Council in managing the growth and development of the precinct. This has not always been the case everywhere else (Committee of Vice-Chancellors and Principals, 1994).

In a sense, the very concept of a Higher Education Precinct is a testament to this desire to achieve a constructive long-term relationship between the City Council and the city's major higher education institutions. Much of this dates back to the 1960s and a consultancy study of that time (Wilson and Womersley, 1967), which resulted in a series of agreements between the various parties about what was being attempted in that area and also in joint formal meeting arrangements which still persist. The wonderfully titled 'Joint Committee for the Comprehensive Planning of an Education Precinct' meets annually at very senior levels from the participating institutions to review progress and to dine, not always necessarily in that order of importance. The major report on the agenda each year is a review of what the City Council has been doing that impacts upon the precinct, usually in the joint names of the City Planning Officer and the City Engineer and Surveyor. This has in turn spawned a series of less formal working relationships which seek to take forward individual issues as they arise. These processes have tended to wax and to wane according to how much development activity was envisaged in the precinct, and recent years have seen a resurgence of this as several major projects have been under consideration. By the early 1990s all parties had concluded that a shared framework for action, which would enable the spending plans of each institution to be co-ordinated in order to improve the precinct both environmentally and functionally, would be appropriate. Austin-Smith:Lord were commissioned to undertake a study. The costs of this were shared between the major educational and hospital institutions, the Central Manchester Development Corporation (since the northern part of the precinct falls into its area) and the City Council. The product of this study (Austin-Smith:Lord/JMP, 1994), with its emphasis on environmental works, on the creation of a greater sense of place, on sign-posting and other related elements, and on access to and mobility in the precinct, quickly began to influence the budgets of some of the participants. The robustness of these relationships was such that they could survive the sorts of difficulties experienced between the City Council and Manchester Metropolitan University over the Stretford Road issue (see Chapter 7). What this appeared to show was that there was a common understanding between participants that the things which united us were much more important than the things which divided us.

In a very important sense, these relationships were two-way. Just as the City Council felt that the actual and the potential roles of the precinct institutions in the economic life of the city were important, so was the role of Manchester as a place where young people wanted to be in selling the city's universities to students as places to come and study. The bases that young people use to make decisions about these sorts of matters are many and varied, but as well as looking at the reputations of courses and of academic staff, young people also think about the sorts of life-styles that they will be able to enjoy during a protracted period as a student in a particular location. Manchester is a selling point for the universities as far as young people are concerned, because it is seen as a very lively place with a thriving youth culture (see below), and the location of the precinct right on the edge of the city centre reinforces this image. Because of this, Manchester University and Manchester Metropolitan University as large institutions tend to be among the highest attractors of student applications for degree courses of all British universities. UMIST is a much smaller institution, and so its absolute application levels are not as high. 1994 UCAS data have been accessed to construct Table 6.2, which shows which universities were the national 'top 8' in terms of numbers of applications for places on degree courses starting that year.

Table 6.2 1994 university applications: the national 'top 8'

1) Leeds University	53,655
2) Manchester University	53,414
3) Birmingham University	51,235
4) Nottingham University	49,184
5) Sheffield University	47,948
6) Sheffield Hallam University	41,173
7) Liverpool John Moores University	40,680
8) Manchester Metropolitan University	40,503

While obviously these are some of the largest higher education institutions in the country, and therefore would be expected to have high absolute numbers of applicants, the table does illustrate vividly the 'pulling power' of large cities for young people. If the constituent elements of London University had been reported as if they were part of one institution, rather than separately, this effect would undoubtedly have been magnified still further.

At the same time as the city's image worked to the advantage of the universities as far as young people were concerned, however, so the city's adverse image as a place of violence, crime and insecurity tended to work to their disadvantage as far as the parents of potential students were concerned. The universities were becoming increasingly conscious of this as an issue, not merely because of parents' concerns, as the 1990s wore on. This produced a degree of tension between the various parties about the appropriate physical response to issues of security (see Chapter 7). Overall, however, the City Council and the universities acknowledged a degree of mutual dependence which argued strongly for working together and building ever-closer relationships, and this approach was the dominating influence in terms of planning actions affecting the Higher Education Precinct throughout my period as head of the planning service in Manchester.

The city's transportation nodality

Chapter 8 contains quite a full discussion of issues surrounding transportation policy, and much of this is directly relevant to the consideration here of transportation nodality as an important part of the city's economic assets. There is, of course, a very long history to this. Developments in both canal and railway building were very major dimensions of the city's rise to prominence during the Industrial Revolution (Briggs, 1982, pp. 88–138). In the 20th century, as well as the city's place in the main-line railway system, important developments have taken place in relation to the national motorway network and in the development of Manchester Airport. This is the primary air gateway for northern England and the country's third largest airport in terms of passenger throughput (likely to be the second largest over the next few years if the second runway proposal is approved, overtaking Gatwick). As well as being large employers in their own right, these transportation industries are vital to the economy of the city in terms of their ability to move people and goods both locally and throughout the world. In addition, transport nodality is an attraction in its own right for certain forms of development, for example, Manchester Airport and Victoria Station (see Chapter 8). The planning job has thus been to ensure that transportation systems are able to function as effectively as possible, that their termini are attractive and accessible, and that where appropriate this nodality is capitalised upon in development terms. Until relatively recently, these tasks, although complex, have not been widely regarded as being very controversial, but the difficulties over the past decade with making local public transport effective (see Chapter 8) and more recently the wider perception of the role transportation considerations ought to play in the debate about urban sustainability (see Chapter 9) have quite considerably changed this view. Much of this affects the city centre, because of its concentration of transport facilities and its role as a major transport destination.

The range of high-technology activities in the city

High-technology activities, although undergoing continuous change, have survived the economic difficulties of recent times much better than have the traditional manufacturing activities which were Manchester's industrial heritage in the immediate postwar period. The city where the world's first computer was put together (in the University of Manchester) has not benefited as much as many other cities from the explosion in economic activity generated by the high-technology revolution. This is an example of the difficulties that can be experienced in hoping to benefit in the local economy from the research that is carried out in a local university. Manchester does have many companies that are active in high-technology industries of various kinds. The City Council has regarded it as particularly important that their operating circumstances should be assisted by (or at any rate not obstructed by) the council's activities as deliverer of local services and as guardian of the local environment. This has tended to lead to fairly good linkages being developed on a personal basis between key people in those companies, and senior Council personnel, both at officer and at elected member levels, in the belief that problems can be solved and opportunities can be grasped via informal mechanisms (Peck and Tickell, 1995). The Chief Executive and the Chief Executive's Department in Manchester have tended to have a particularly important role, pulling people from relevant departments together either to tackle difficulties or to look proactively at possible development opportunities that may have been raised with them.

While quite a lot of this is inevitably about relating to the companies that the city has already got, sometimes there is an opportunity to acquire a significant piece of inward investment in the high-technology field. This tends to be seen as a matter of very high priority, not just because of the direct economic benefits but also because it is seen as an external statement of confidence in the city. Such an opportunity was available to Manchester in the late 1980s in the form of the search by Siemens for a northern England headquarters location for part of their operation. Siemens chose to locate this activity in Manchester in preference to many other locations that had been lobbying for it. This included many others where the grant availability to support the investment was said to be much better than for the site in south Manchester. My impression from extensive personal involvement in this process was that the factors in Manchester's favour were the location of the site (halfway between the airport and the city centre, and within a few hundred yards of a motorway junction); the fact that the site was itself in an attractive part of the city; the proximity of a range of academic contacts in the city's universities; and the fact that the City Council was prepared to go out of its way to help Siemens with the project. This included the arrangements for the acquisition of the site, which the council owned.

This site was one which the council had been reserving for a major piece of inward investment. Until then however, the site was intended to remain in use as educational playing fields, although an embryonic relocation plan existed to enable early access to the site to be given. What this actually meant in local planning terms, however, was that it was a green piece of land that people had been used to having access to for formal and informal recreational purposes, in a part of the city where there were some very active community groups likely to hold and to express strong views about development proposals in that area. Between the company and ourselves, it was quickly agreed that while the council would have to handle the arguments that might arise about the loss of the site to development, both our interests would best be served by producing a building of quality; and by a very open local consultative process which gave people opportunities to be involved in discussions and to be satisfied about the standard of development being achieved.

As a result of this local people were on the whole prepared to accept that the site did have a contribution to make in securing a headquarters investment by a high-technology company, but only if the fine words about quality were translated into reality. As Assistant City Planning Officer responsible for the Area Planning Division at that time, I headed this process from the council's side, and the willingness of Siemens' shadow management for the new complex to keep working with us to seek to improve the scheme, even when a set of the necessary permissions was already in place, meant that we were able to secure for the site what I believe is one of the best modern buildings in Manchester. Something like this is always a local talking point, and that is I believe a good thing; it is better to produce buildings that excite interest and attract comment (even if negative) than to go for the safe, the boring or the mediocre. So there are varying views in Manchester about the Siemens building, although on the whole most of the feedback I have had has been supportive rather than critical.

A fragment of this negotiation process was captured by Professor Patsy Healey of the University of Newcastle upon Tyne (Healey, 1992a), because she spent a day with me in March 1988 shadowing what I was doing, to look at the processes of communication between a planner and the customers of the planning process. The exchanges recorded in that article were typical of a process that extended over a considerable period; trying to balance the desire of the project architect and the company to produce a high-quality building which conveyed the sorts of messages that they wanted to communicate, and our desire not merely to ensure that a good building was achieved for Manchester but also that known local views were taken into account and that the most expeditious routes through the formal processes of decision-making could be found.

Technology does not stand still. At the time of writing, Nynex have won the local cable television franchise and are busy undertaking cabling work throughout the city. This is another example of an activity which generates a degree of environmental disruption at first (for the most part outside formal planning control) but produces technological benefits once this has been done. It will be interesting as this process unfolds to see what the effects of all this will be on the city's economy, but one of the early effects has been the involvement of the company with the new indoor arena (the Nynex Arena) at Victoria Station. This perhaps illustrates the extent to which investment in high-technology activity can penetrate and reinforce investments in other aspects of the city's life and its economy.

The city's cultural and sporting life

Until recently, it certainly would not have been usual to include in a list of a city's primary economic assets its cultural and sporting life. However, it is not only that these sectors are important generators of jobs in their own right but they also play a major part in creating a dynamic and an exciting image of the city as the place to be. Several other western European cities have clearly had similar experiences (Bianchini and Parkinson, 1993). Perhaps particularly, these sectors contribute considerably to the development of the city's visitor economy, described above. They are thus a very important part of the process of trying to 'sell' a more positive image of the city (Kearns and Philo, 1993).

Some aspects of the city's cultural life are long-standing. For example, Manchester City Art Gallery and the Central Library have long been regarded as major facilities of regional significance, providing services for people from a wide area. Since both are local authority facilities, they illustrate the well-known phenomenon that the local taxpayers of core cities are paying to provide services not just for themselves but also for a much wider community, and we have not yet found a wholly satisfactory solution in Britain to the problem of equitable funding for these sorts of activities. Similarly, Manchester for a long time has had a reputation in the field of classical music, with the Hallé Orchestra as the main standard-bearer for many years. The City's role in the media industries is also now well established with television and local radio adding to the role that the newspaper-publishing industry has long played.

Some aspects of the city's cultural life, however, are much more recent. The renaissance of theatre in the city is something that has taken place in the past 20 years or even less, with major projects to create the Royal Exchange theatre-in-the-round and to reopen the Palace and the Opera House. Fashions in popular music tend to come and go, but in the late 1980s and early 1990s particularly, Manchester was seen as one of the country's pop music capitals both in terms of bands and of venues. The story is often told of the potential university student coming for interview, whom when asked whether he has any questions immediately asks for directions to the Hacienda, one of the city's best-known clubs.

The regeneration of the Castlefield area of the city centre from the late 1970s was partly accomplished through a major increase in museum provision in the form of the Greater Manchester Museum of Science and Industry and the Air and Space Museum, both of which involved the extensive reuse of run-down historic buildings. There was also the creation by Granada Television of the Granada Studios Tour. The construction of a major exhibition venue in the early 1980s (the G-MEX centre) out of the former Central Station, closed at the time of the Beeching cuts of the 1960s, is another instance of similar processes at work. Many other examples could also be found, particularly of the flowering of the arts at a more local level.

In terms of its impact upon the planning process, much of this has involved work on major projects, sometimes over a protracted period of time. Recent examples have included the project to construct a new concert hall on Lower Mosley Street, and the project to create an extension to the City Art Gallery. Both of these involved competitions with design elements to them, in which the Planning Department was extensively involved. These sorts of activities are usually accompanied by, and sometimes led by, the need to generate funding packages which enable good-quality schemes to be built, and the main role in this work is typically taken by officers from the Chief Executive's Department. Public–private partnerships are usually involved with public resources being used to lever in private funds, often with models that have to be reconstructed each time because of the particular circumstances of each case.

The planning role in these situations is as part of a corporate team, contributing both to scheme generation and to evaluation by a variety of means including brief-writing, information provision, project assessment in relation to planning criteria, and scheme negotiation as part of the development control process. It is important to stress that the council has approached these issues in recent times on the basis of wanting the best schemes that it can get. It is not for the planning service in these sorts of circumstances to set standards that turn out to be unattainable and then to stick to them in a way which obstructs the achievement of major projects. This pragmatism doesn't always go down well in all quarters, but the willingness to see problems as there to be solved, including if need be with the best compromise possible, has contributed significantly to the achievement of several major projects in recent years.

The same could be said about the development of sport in the city. Manchester's bids for the Olympic Games of 1996 and 2000, and its ultimately successful bid for the Commonwealth Games of 2002, have probably done more to raise the city's profile than any other single activity. I have written extensively about this elsewhere (Kitchen, 1993c; 1993d; 1996f) and so readers who want to follow this up in more detail can readily do so as well as by looking at the main submitted bid documents (Manchester Olympic Bid Committee, 1993; Manchester Commonwealth Bid Committee, 1995). The main Planning Department contributions to these processes were as follows:

- Helping in the first place with the basic development work on the venues strategy. This was a much stronger part of the 2000 bid process that it was of the 1996 bid process, arising from the determination that the City of Manchester itself would get more direct regenerative benefits from a bid which carried its name (and which it would have to underwrite) than would have been the case with the 1996 bid. This involved both a strategic level of activity to try to determine in several fields what the options were for locating facilities in the city, and then a more local assessment of the best ways of achieving this once strategic decisions had been taken. Both the stadium/velodrome package at Eastlands and the arena package at Victoria Station went through stages of activity of this type, with the staff involved varying according to the tasks to be carried out but with a strong input throughout from the appropriate area teams. In both Eastlands and Victoria Station, there were competitive elements; for the stadium in conjunction with an Eastlands master plan. Relevant staff made a large contribution to the process of assessing competition submissions and I myself was involved in an advisory capacity in the judging process.
- Working closely with the project teams established to take forward the major pieces of infrastructure work. The Victoria Station arena project was largely conceived through a process of the Planning Department working from the late 1980s in informal partnership with the Inter City Property Group and their retained architects (Austin-Smith:Lord). In addition, a senior member of staff from the department's north area team was seconded to the corporate team assembled to deal with the Eastlands site assembly and clearance project for its duration. The department also played a major role in the work on the Olympic Village, from the decision over its location (see Figure 6.4) to much more detailed work on layout issues and principles. As well as work on the form that would be taken for the Games themselves, this area of activity also included work on environmental improvements that could be undertaken relatively quickly to improve the impressions given to visitors coming to the city as part of the bidding process.
- Helping with presentations of various kinds to the stream of visitors received by the city as part of the bidding process, and helping with site visits. This was on quite a large scale; for example, about 80 International

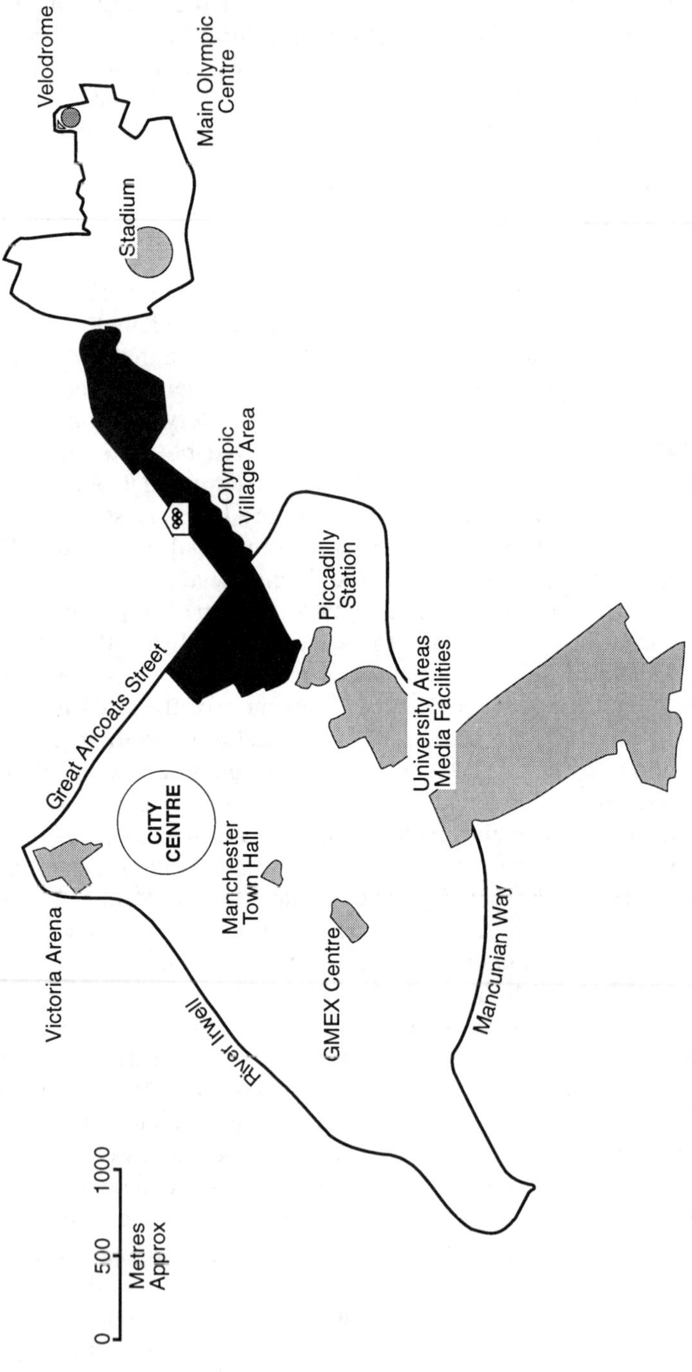

Figure 6.4 The Olympic Village location in relation to key games sites

Olympic Committee members came to Manchester as part of the 2000 bid process, often in small groups or alone, and for those who had the time to go out on site and look for themselves, departmental staff were often involved in showing them around and answering questions informally as part of such activities.

- In my particular case, acting as spokesman at conferences in Britain for the Olympic bid process, as part of the job of winning 'hearts and minds' in the UK. Between Autumn 1992 and September 1993 I did this on an average of roughly one a month, showing that conference organisers thought the Manchester Olympic bid a topical and interesting subject. We always accepted these invitations if we could because we believed that it was important that we should be willing to go out and talk to people 'at home' about what we were doing in their name (because the Manchester bid was also the British bid). We also hoped that this might help to counteract some of the adverse stories emanating from some branches of the London media.
- Talking to organisations whose support we needed. This was largely organised on the basis of who had the best contacts in the first place, so that for example I played a leading role in presentations to English Heritage and to the Royal Fine Art Commission because we wanted to be able to claim their support (which was forthcoming) in a publication about the bid's environmental credentials.
- Participating in the ongoing work about some of the key topics, where the relationships between decisions about venues and forms of development and decisions about how to handle those topics were essentially iterative. The best example of this is probably in relation to the topic of transport, where the approach was constantly evolving, the vital task being the movement of competitors, officials, the media and spectators around the city without causing major difficulties either for them or for the city itself. The Planning Department's Strategy Group, by virtue of its transport expertise, made a more or less continuous contribution to the processes of thinking about and of testing options for what was one of the most difficult problems to be tackled in Games planning.

Particularly during the 12 months immediately preceding the decision in September 1993, there was absolutely no doubt that as far as the council's political leadership was concerned the Olympic bid was priority number one. This fed into the work of departments via a series of imperatives from the Chief Executive's Department, very often with short deadlines against them and sometimes with tempers that were almost equally as short. I think it is fair to say that for most of the staff for most of the time, this was accepted as being a result of the very particular situation that the city was in; and the excitement of playing a part in the process (which was not, of course, an opportunity that most staff in most planning departments would ever experience) was a very real compensation for the difficult working circumstances that sometimes arose.

Some issues became unquestionable during this period. For example, we were not able to deal very effectively in the work on the Olympic Village, with emerging issues in terms of listed buildings and the possible promotion of a conservation area, on the back of a significant number of proposed further listings by English Heritage of canal-related architecture along the Ashton Canal, because this would have meant reconsidering some statements about the village that had already been published as part of the bid documents. Ironically, it might have been possible to have used such a process to lever in more external resources which could have helped with the construction of the village, and would almost certainly have generated a still more fulsome statement of support from English Heritage. The subject, however, was simply regarded as being taboo, as we learned the hard way when we tried to raise it. None the less, the overall reaction of the departmental staff who had participated in the process in one way or another was positive, and the real sense of disappointment in Manchester at large, when the 2000 Games were awarded to Sydney, was shared by the department's staff.

The city's people

Putting the people of the city last in this inventory is not to suggest that they are less important than the other elements; after all, enterprise and initiative have to be exercised by individuals if anything is to materialise. The point here is that, throughout the history of the city, in many ways its people have always been a primary economic asset; as labour force, as source of ideas, as ambassadors, as purchasers of other people's goods and services, and in many other ways. This has included not only people whose family histories go back a long way in terms of the city, but also people who have come to Manchester from all over the world and often under very difficult circumstances. In the 19th century, Manchester's reputation as a radical city made it one of the natural destinations for people who left their own homes in other parts of Europe often after suffering persecution of various kinds, but they integrated into the city's life and made their own distinctive contributions (Briggs, 1982). This process has continued in the 20th century, again with beneficial effects. Two examples are the creation of Chinatown in the city centre, and the development in Rusholme of a major concentration of Asian restaurants and shops. Both of these have brought investment on a large scale into areas of the city which were quite run down; and both these areas have now become tourist attractions in their own right adding to the range and the quality of eating-out facilities in the city.

Perhaps the most important response the planning process has had to make to these developments has simply been not to get in their way. For example, Chinese businesses sometimes want to put advertising boards on the Victorian and Edwardian buildings that constitute most of Chinatown, in ways that might not always be consistent with views about how such buildings in a conservation area should be treated. But, within reason, how much does this really matter? If this sort of activity helps to give an authentic Chinese feel to the area, and helps old buildings to stay in use or to be brought back into use, then the willingness to be relaxed about advertising boards may be a relatively small price to pay; and it certainly helps to convey to Chinese business people the impression that the planning service is there to be helpful rather than to be bureaucratic and obstructive. The other dimension to this, of course, is that it is an important element in the planning service's philosophy of working closely with its customers, and working with local communities on an area basis.

Conclusions

A sound economic base is vital to the future of the city. A city without a sound economic base is a city which is in decline, and while most cities have periods of decline at some points in their lives there is always a risk that the longer and the steeper is the process of decline the harder it can be in practice to pull out of it. Political leaderships in large cities tend because of this to give a very high priority to sustaining and to improving the economic bases of their cities, and as this chapter has shown Manchester is no exception to this. The approach that has been taken for some time now is to try to understand what the city's actual and potential strengths are in economic terms, and then as far as possible to build on them.

The story told in this chapter is of a city trying to come to terms with its role in the late 20th century in a changing world. The smokestack economy epitomised by Manchester's rise to prominence during the Industrial Revolution has gone, and will not return. The things that are the city's strengths approaching the millennium are really the things that contemporary cities need to be good at, which are to do with their ability to attract large numbers of people for particular purposes. Instead of the fight to retain and attract manufacturing industry, the emphasis is on transport nodality, on knowledge industries of all kinds and on activities like the arts and sport which simply would not have featured at all in an economic assessment 20 years ago.

The planning process in all this should be an enabler. Planning actions can contribute significantly to creating the circumstances whereby projects can be successful, and planners as members of teams can play a variety of roles working alongside people from many other disciplines in helping to make this happen. This is a team job, and planners need to see what they are doing as fitting in with and reinforcing the contribution of others to an agreed corporate end, rather than as a series of hurdles set out for a separate race. Planning also deals with the very direct and the very local consequences of much of this, however; and it often does this in a consultative style where there is (and is seen to be) some balance being struck between the wider economic interests being served and the immediate local impact of the project. Under these circumstances it is very easy for the planning process to appear unhelpful and obstructive, either to the senior politician who has already decided that the wider good must prevail, or to the local citizen who believes that he or she is faced with a *fait accompli*. My experience is that people do sometimes jump into these stereotyped roles all too easily. When this happens, planning staff can feel caught between two conflicting sets of pressures with no obvious way out, save that of trying to be as honest and as open with people as possible about the situation and the procedural steps to bring it to a conclusion, and then of trying to give the best advice in those circumstances. This may mean looking for ameliorating actions that can reduce the local impact of a project, even though it may be clear that the project itself will proceed, and this can add to project costs and cause difficulties. But if the planning process is indeed to strike a reasonable balance between the long-term interests of a city and its people who stand to gain from a project once it is operational, and the interests of those people who will have to live with its immediate impact including during its construction phase, it must tackle issues like this.

Two particular dimensions of some of these considerations tend to surface fairly frequently in political debates:

- How over-riding a priority is economic development, and for whom? How effectively does the local political process resolve conflicts and tensions between economic development initiatives and other policy concerns (for example, in terms of environmental impact), when the council's political leadership is probably already lined up behind the economic development approach?
- How important are major or 'prestige' projects, which can consume time, financial resources and political attention, compared with the 'bread and butter' and usually rather unglamorous process of delivering local government services, which have been under considerable pressure anyway because of the financial circumstances of local government? There are real questions to be asked about who actually benefits from some prestige projects, and how this relates to the needs of some of the poorer sectors of our urban communities that are looked at in more detail in Chapter 7 (Loftman and Nevin, 1995).

What lies behind these questions is a growing recognition that work on a city's economic base and work on tackling the economic difficulties of its most deprived citizens cannot be two separate things. Work to ensure as far as possible that a city's economic base continues to evolve positively is clearly necessary but there are real doubts about whether this by itself ensures that the benefits of these processes are felt by the most deprived sectors of the city's population. Chapter 7 explores this proposition in more detail from the other end of the telescope, compared with the economic base perspective adopted in this chapter.

Even leaving aside the awkward question of who actually benefits from economic base work, there are no easy answers to the sorts of questions that have been asked above, although people can be quite quick to take sides in individual circumstances. The political process takes decisions about matters such as these, it is hoped from as well informed a perspective as can be achieved. But whatever decisions are taken the basic problems remain of having to make choices between often incompatible objectives, and of having to prioritise actions with inadequate resource levels to tackle everything needing to be done. These sorts of tensions are visible in many of the key areas of local government decision-making that have to be faced today in cities like Manchester and no doubt in many other places as well, and they will resurface in the chapters which follow. These tensions are also very visible when looking at the planning job in Manchester. Because of the things it deals with, it may be particularly susceptible to being in the limelight (or in the firing-line, depending upon one's perspective and language) when these tensions surface and have to be resolved, particularly because the requirements and traditions of public participation in planning are not part of many other forms of economic decision-making by councils. In particular, planning can be left with its familiar job of balancing issues when some of the elements of that balancing process are already politically committed. This phenomenon too will be illustrated on several occasions in the subsequent chapters.

7

Planning in the inner city

Introduction

Many people's image of Manchester, as the core city of one of Britain's largest conurbations, may well be that it is in effect all an inner city. Like a lot of images, it is clear that there is some substance to this, and yet at the same time 'the inner city' as a concept has a more specific meaning than this. The first part of this chapter, therefore, discusses this concept in its wider context. The chapter proceeds to look at inner-city policy from the top down, as it has been conceptualised and applied by the government, which includes a case study of Central Manchester Development Corporation as an imposed top-down mechanism; and then it looks at inner-city policy from the bottom up, as it has been thought about and attempted locally. Two particular examples of these approaches are then examined in more detail; the City Pride process which looked at the area as a whole in a broad context, and the regeneration of Hulme through the City Challenge process. Finally some threads are pulled together in terms of outstanding issues. Throughout, specific comments are made about particular planning dimensions that have arisen and about the particular contributions made by Planning Department staff. It is important to emphasise, however, that much of what has been said earlier in this book about the basics of the planning service applies to the inner city just as much as to other parts of the city. In particular, it was the experiences of planning in the inner city that shaped Manchester's approaches to plan-making that have been described in Chapter 4 (see also Fudge and Healey, 1984), and these points need to be remembered when reading this chapter.

By its nature, this is a small and highly selective slice through a large subject, but it is hoped this specific commentary on inner-city and planning issues in Manchester will complement some of the literature already available about the generality of inner-city policy (for example, Lawless, 1989; Keith and Rogers, 1991; Ambrose, 1994; Atkinson and Moon, 1994; Robson, 1988; 1994; Hambleton and Thomas, 1995; Harding and Garside, in Stewart and Stoker, 1995, pp. 166–87). For a specific discussion of the place of local

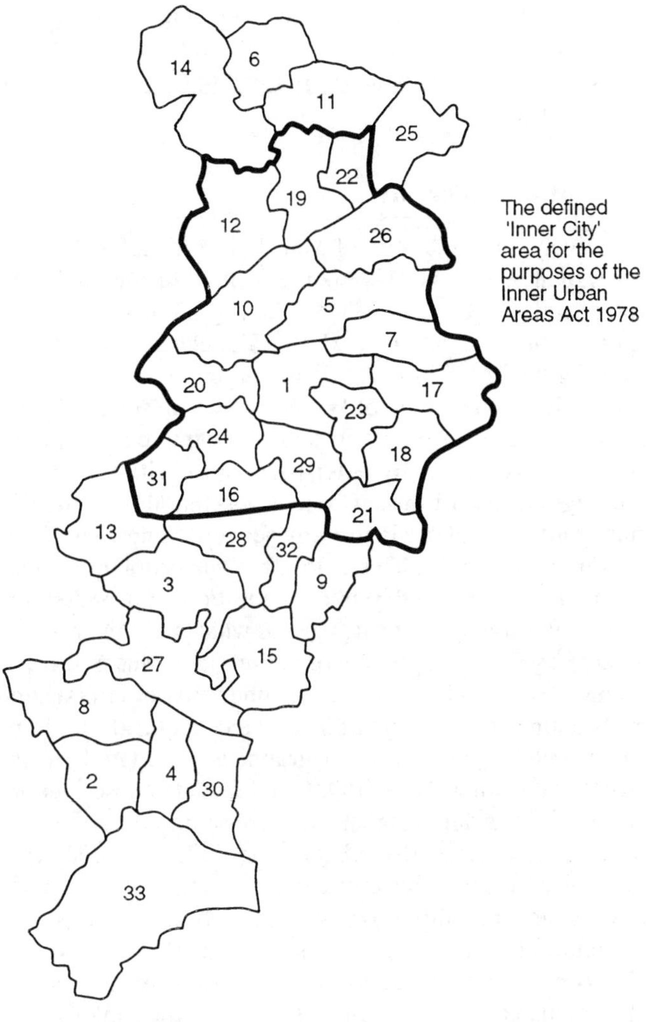

City of Manchester Wards

1. Ardwick
2. Baguley
3. Barlow Moor
4. Benchill
5. Beswick & Clayton
6. Blackley
7. Bradford
8. Brooklands
9. Burnage
10. Central
11. Charlestown
12. Cheetham
13. Chorlton
14. Crumpsall
15. Didsbury
16. Fallowfield
17. Gorton North
18. Gorton South
19. Harpurhey
20. Hulme
21. Levenshulme
22. Lightbowne
23. Longsight
24. Moss Side
25. Moston
26. Newton Heath
27. Northenden
28. Old Moat
29. Rusholme
30. Sharston
31. Whalley Range
32. Withington
33. Woodhouse Park

Figure 7.1 Manchester's inner city

government in the urban regeneration process, see Audit Commission (1989). More generally, Robson *et al.* (1994) have produced a major piece of work evaluating what urban policy is actually achieving, focusing on Greater Manchester, Merseyside and Tyneside. In addition, for a general discussion about the problems of the people who live in the inner cities of both Britain and the USA, see Jacobs (1992).

The nature of the inner city

Administratively, the inner city of Manchester as defined for the purposes of the Inner Urban Areas Act 1978 consists of 17 of the city's 33 wards, or just over half in total. Only a small part of the city on its northern side is excluded from this area, but a large part of South Manchester does not have inner-city status (see Figure 7.1). Thus, the defined area includes the whole of the city centre but excludes the whole of Wythenshawe, despite several attempts over the years to argue that the social and economic circumstances of Wythenshawe make it very like many other parts of the city that do have inner-area status. The significance of this, of course, was that during the life of the Urban Programme and particularly in its early years funding went to the defined area but not to areas outside it. This is the familiar problem of needing to define areas to enable the day-to-day implementation of a policy initiative to proceed, but these examples of what is in and what is not in perhaps serve to show the inevitable crudity of a process of definition of this nature.

The thing that does characterise the inner city, of course, and that was the basis for its definition in the first place, is a concentration of problems. In the terms of the 1990s, this has been graphically updated by the Manchester Health for All Working Party (1993), and it is also described in Robson *et al.* (1994, pp. 215–32). A large statistical exercise based upon this material could be undertaken to look at the relative distribution of deprivation within the inner city, between the inner city and the rest of the city, and then with the Greater Manchester conurbation as a whole. In the interests of simplicity, the main differences that would emerge from such a comparison are summarised in Box 7.1. A more detailed analysis of the spectrum of deprivation within the inner city can also be undertaken by looking at the rank orders by clusters of inner-city wards against the following 10 indicators in the 1993 Manchester Health for All Working Party study:

- Percentage of total residents aged 0–15, from the 1991 Census.
- Percentage of households headed by a single parent, from the 1991 Census.
- Percentage of unemployed residents aged 15–59/64, from the 1991 Census.
- Percentage of male residents aged 16 or over in Social Class V (unskilled), from the 1991 Census 10 per cent sample.
- Percentage of households without a car, from the 1991 Census.
- Percentage of households receiving housing benefit in March 1992, according to the Manchester Welfare Benefits System.

Box 7.1 The inner city in comparison with the rest of Manchester and the whole conurbation

- The population of the inner city is generally younger than that of the city and of the wider area, with both lower numbers of residents of retirement age and higher numbers of children aged 15 or under.
- The inner city has a lower number of single-pensioner households than does the city as a whole, but a number equal to that in the wider area. Thus, it is the outer areas of Manchester and not the inner city which contains the greater concentration of single-pensioner households, which is perhaps not many people's typical model of the inner city. This indicator is often thought of in terms of social policy as containing one of the significant categories of people 'at risk' and likely to need social services support.
- The inner city does have a significantly larger share of households headed by a single parent than does the city as a whole, and as far as the rest of Greater Manchester is concerned the inner city has more than double the figure for the wider area.
- The inner city's figure for households lacking basic amenities is slightly lower than that for the city as a whole, but both are approximately 50 per cent greater than that for the rest of Greater Manchester. What this shows is that on one of the traditional measures of housing quality the city's housing stock is in significantly worse condition than that of the surrounding area, but that within the city this is actually slightly more of a problem outside the inner-city area than it is within it.
- The inner city has a greater concentration of unskilled males than does the city as a whole, and in turn the city has a significantly higher score against this indicator than does the rest of Greater Manchester.
- More than 9 in 15 inner-city households don't own a car, whereas for the city as a whole the figure is about 8 in 15 and for the rest of Greater Manchester it is less than 5 in 15. Thus, the inner city has a higher score than for the city as a whole, but of much greater significance is the large difference between Manchester and the rest of Greater Manchester. This is often regarded as being not only an indicator about transportation and mobility (since households without access to a car are usually reliant on the public transport system) but also an indicator of relative affluence or poverty; and if taken as such it would show that inner-city residents are substantially poorer than those of the rest of Greater Manchester and substantially more dependent upon the public transport system. The implications of this latter point in terms of transportation policy are discussed in more detail in Chapter 8.

- Fertility rates are progressively higher in the city as a whole and then in the inner city as compared with the whole of the North West Regional Health Authority area. This is not unexpected in the light of the figures which show the inner city to have a relatively young population.
- Low birth weights are slightly more of a problem in the inner-city area than in the city as a whole, but are noticeably more of a problem than in the region as a whole. As well as being an indicator which has important things to say about a segment of health-care provision, this is often regarded also as being quite closely related to relative poverty.
- The standard mortality ratio figures for males under 65 are substantially worse in Manchester than they are in the region as a whole, and then significantly worse in the inner-city area than in the city as a whole. This suggests that Manchester in general, and the inner city in particular, are relatively unhealthy places to live, although the factors that go together to make up a judgement such as this are complex in the extreme.

- Percentage of low-weight births in 1991, from OPCS Birth Tapes.
- Standard mortality ratio for males under 65, from OPCS Death Tapes and the 1991 Census.
- Percentage of households living in houses lacking basic amenities, from the 1991 Census.
- Percentage of households renting their house from the City Council, from the 1991 Census.

With the highest numerical score in each case being recorded as rank order 1 and the lowest numerical score being recorded as rank order 5, the five clusters of inner-city wards defined for the purposes of the 1993 study can be seen as forming a spectrum of deprivation as follows:

Central A	7 rank order 1s (relatively most deprived)
North C	8 rank order 2s
North B	7 rank order 3s
Central B	5 rank order 4s
Central C	8 rank order 5s (relatively least deprived)

This spectrum at the level of the wards contained within each cluster is shown in Figure 7.2.[1] This suggests that the most deprived wards are those to be found in a ring immediately around the city centre (which is the southern part of Central ward), with relative deprivation declining the further away from the city centre one goes. This analysis could be repeated at individual ward level or even lower, and it would show the pattern to be more complex than this although it would not fundamentally change this conclusion. The physical characteristic that best defines the wards that fall in the most deprived clusters on this basis is that they are typified by housing redevelopment of the 1950s/1970s period, including much of the city's system-built high-rise accommodation from some of the later years of this period. This understanding of the nature of the inner city is fundamental to an understanding of what has been attempted there and why, and it therefore shapes the remainder of the material in this chapter. However, it is important to remember that analytical expressions about the components and the distribution of deprivation mask a great deal of individual human suffering and loss of feelings of worth; for a sociological analysis of this in Manchester and Sheffield, see Taylor, Evans and Frazer (1996, pp. 163–79).

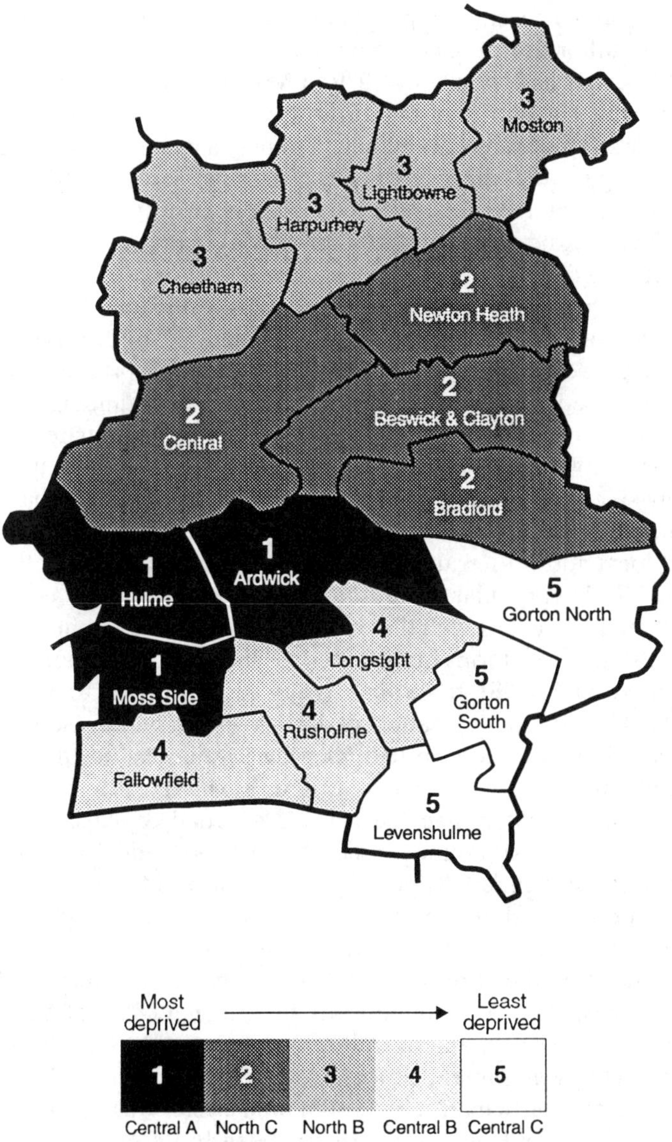

Figure 7.2 The spectrum of deprivation in the inner city

Policy approaches to the inner city from the top down

My general views about the perspectives that successive governments had adopted towards inner-city policy by the time of the mid-1980s are wholly given away by the title of a book chapter I produced at that time: 'Inner city policy and practice 1975–1985: reflections on a lost opportunity' (Kitchen in Willis, 1986). This looked at inner-city policy from the perspective of someone who had been deeply involved in the preparation of the first inner-area programme for South Tyneside (a 'programme' authority, which was the second tier down of authorities to benefit from the Inner Urban Areas Act 1978) in the late 1970s, and had moved by 1979 to Manchester (a 'partnership' authority, which was the top tier of authorities to benefit and those that received the largest amounts of money) and begun to play a part in its inner-city work. This chapter argued that inner-city policy as far as government was concerned was characterised by the mid-1980s by a process of drift. The idea of a programme which looked at the needs of an area over the long term, identified the resources needed to begin to tackle these needs over a sustained period, and sought not merely to utilise extra resources to this end but primarily to 'bend' the mainstream programmes of central and local government had for all practical purposes disappeared before its teething problems had been seriously addressed. Instead, what we had was a free-standing programme which had developed its own bureaucracy, which was inadequately resourced to tackle the problems and in any event carried no guarantee of its continuation for any significant period of time, and which involved the abandonment of the concept of 'bending'. The government's policy towards the spending regimes of local authorities in general and what it regarded as left-wing Labour authorities in particular meant that it was taking far more away with the one hand than it was giving with the other. To illustrate that point, I showed in that chapter that in real terms Manchester's Rate Support Grant had fallen by £30 millions from 1980/81 to 1984/85 (from £120 millions to £90 millions) whereas its share of the allocation of Urban Programme resources to the Manchester/Salford Partnership had risen by £3 millions on this same basis (from £3 millions in 1980/81 to £6 millions in 1984/85). Thus, the losses were roughly ten times greater than the gains. My broad conclusion was that a worthy initiative was being undermined by recession, ideology and neglect; far from things in the inner city getting better as a result of nearly a decade's worth of policy application, they actually appeared to be getting worse.

Since that time, we have had numerous often short-lived policy initiatives, often introduced at the drop of a minister. Box 7.2 summarises the most significant of these. It is very doubtful whether the net effect of all this policy overload was beneficial in any overall sense, if that is measured in terms of what all this activity was intended to achieve (Robson *et al.*, 1994). Certainly my perspective as a participant-observer in Manchester was that the incidence of many inner-city problems was worse by the early 1990s than it had been a decade earlier, and this appears to be supported by recent research not just in

Box 7.2 Major central government policy initiatives towards inner cities: late 1970s/mid 1990s

1978	Inner-area partnerships and programmes (57 authorities designated under the Inner Urban Areas Act 1978).
Early 1980s	Declaration of 30 enterprise zones.
1981–92	Declaration of 13 urban development corporations.
1982–88	various grant regimes established: • Urban development grants in 1982. • Urban regeneration grants in 1987. • City grants in 1988.
1985	City action teams established, taken further in 1986 with the creation of task forces.
Late 1980s	Estate Action programme commenced, targeting 'difficult to let' and 'priority' housing estates.
Late 1980s/ early 1990s	Major educational and training changes, such as the creation of training and enterprise councils winding-down of the Urban Programme.
1992	City Challenge process started (two major rounds, and then incorporated into Single Regeneration Budget).
1993	City Pride invitations to London, Birmingham and Manchester.
1993	English Partnerships established.
1994	Integrated regional offices of government established, along with Single Regeneration Budget.

relation to Manchester but in relation to many other inner areas (Robson, Bradford and Tye, 1995; Lawless, 1996).

The conventional wisdom behind much of this was the idea of property-led urban regeneration, through which trickle-down economic benefits would reach the parts that other policies had failed to reach. For a full discussion of this thinking, see Healey *et al.* (1992), Turok (1992) and Deakin and Edwards (1993). For part of this period, this included a series of attempts to sideline local authorities; they were part of the problem rather than part of the solution; although certainly by the time that the City Pride invitations were issued the representative and the co-ordinating roles of local authorities (at least) had been rediscovered (see the arguments about this in Audit Commission, 1989). It is instructive none the less to think about the extent to which property-led urban regeneration and trickle-down economic theory were likely to be congruent with the primary characteristics of Manchester's inner city identified in the early part of this chapter. It seems to me that it would be very hard to conclude that there is an obvious and direct relationship between these two, given that the major areas of deprivation were themselves the areas that had been the subject of comprehensive redevelopment in previous decades of slum clearance, although certainly some of this legacy was unsuccessful and needed to be tackled. It is perhaps symptomatic of these difficulties that the area in Manchester where its development corporation was established was not any of these areas of greatest deprivation, but rather was the southern one-third of the city centre where at that time hardly anybody lived. Of course there is a case for a property led approach; for example, it seems to me to be appropriate to tackle the problems of a large and declining industrial estate such as that at Trafford Park via the development corporation mechanism. But an emphasis on property-led regeneration to the virtual exclusion of almost anything else for most of the late 1980s and the early 1990s has not addressed many of the fundamental problems of the people of the inner city, which are about their economic, social and environmental circumstances and are only indirectly and partially affected by property redevelopment.

Central Manchester Development Corporation

CMDC's area is shown in Figure 7.3. Quite why this area was thought to be particularly in need of the ministrations of a development corporation was never wholly clear to me, and I suspect I was not alone in this. Robson (1988, pp. 131–32) ventures the view that most of the short-life mini-UDCs announced in 1988 including central Manchester were a combination of problem and opportunity, offering through the structural decline of traditional industry and the availability of land and property development opportunities often alongside waterways the chance over a relatively short period of time to promote a mixture of housing, recreational and commercial development. Whatever the reasons, on the back of its 1987 general election victory, the government had decided that Manchester was going to have a development corporation, and that was not a decision that was regarded as being up for discussion. The government appears to have alighted on the chosen area, which is the southern one-third of the city centre, after looking at alternatives which were more visibly 'inner city' areas but much less promising propositions in which a short-life body could achieve something. Having made the point that it did not feel that the general circumstances of the area warranted such a new policy instrument, therefore, the City Council decided that it would accept the reality of the situation and work with the new organisation. The basis for this was that the new body might well be able to bring into that part of the city resources that it would not otherwise have obtained, and if this was going to happen it was better for the council to be attempting to shape this in ways that it saw as being in the long-term interests of the city, than to stand off from this process.

Really from the outset, therefore, there was not the history in the Manchester case of major friction between the council and the development corporation that has been experienced elsewhere (Imrie and Thomas, 1993). There are four particular components to this:

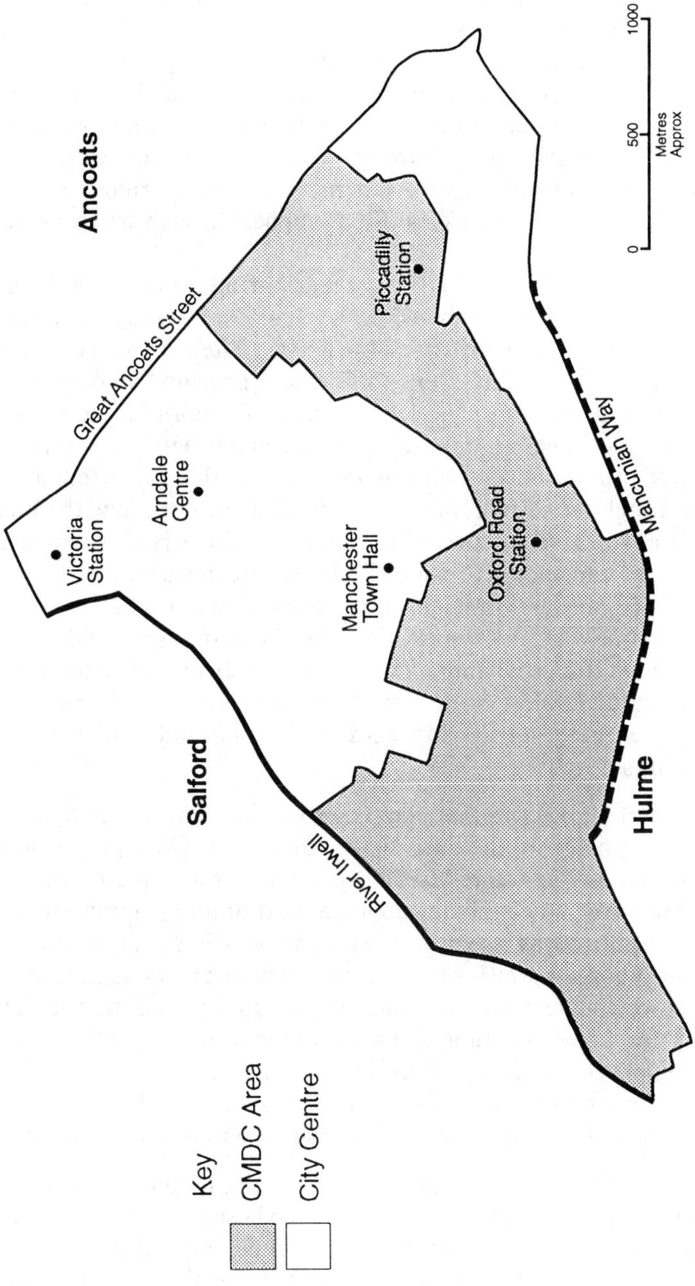

Figure 7.3 The area of Central Manchester Development Corporation

- From the beginning, three members of the City Council (including the Leader of the Council and a senior Conservative member) were amongst the 10 members of the board, and this direct contribution to the membership of the board stayed constant throughout in personnel terms, even surviving the loss by the Conservative member of her council seat in Didsbury.
- Very quickly CMDC decided to adopt the principles of the City Centre Local Plan and make it one of the starting points for its own development strategy; and it also took as a starting point projects that were in the council's own environmental improvement rolling programme so that it could begin very quickly to make a direct impact through environmental investment.
- The existence of a development control agency arrangement with the City Planning Department was combined with a very substantial measure of agreement in practice about what were the appropriate decisions in relation to submitted applications. The agency agreement provided an opportunity for me to attend the board of the Development Corporation and put my point of view if there was an important disagreement between myself and Development Corporation staff on a major case, but this only needed to be used twice throughout CMDC's life. On one of those occasions the board accepted my advice and on the other it didn't, which both sides would probably accept as an even score in contentious circumstances.
- There were no significant strategic differences between CMDC and the City Council about what in essence they were both trying to achieve, and the ability to work in partnership on major initiatives to add value to what each had separately to offer (for example, the new concert hall scheme for Lower Mosley Street, and the Manchester Olympic bid) was seen as a major source of strength.

The consequence of all this was that planning work in Manchester city centre settled down very quickly on the basis that CMDC was now a player on the stage. Provided that no one wanted to discuss whether this was a desirable or a necessary thing, or whether the development corporation's remit might be widened to cover other areas (as happened on one occasion to my knowledge, and possibly on others as well), or any other such contentious matters, an amicable and co-operative relationship developed on the basis that the existence of CMDC for a period of time was a fact of life to be turned to the city's advantage as much as possible. When an objective overview of CMDC's achievements becomes possible once some time has elapsed after they closed down on 31 March 1996, I suspect this will broadly be in four categories:

1) CMDC has played a very valuable role in helping to get major projects started. The new concert hall is undoubtedly the best example of this, since it took a sustained effort between the City Council and CMDC to lever in the public and the private money needed to make this possible and to create for the city a potentially dramatic new space around a new canal basin. For most of this period, the City Council could probably be fairly

described as being in the lead in this process, since amongst many other interests it had in the project it was the landowner. The process none the less, irrespective of who took the lead at particular times, was a co-operative one. At the time of writing, the concert hall itself, the associated office buildings and the canal basin are all under construction, and so it is too early to say what the public reaction to what has been achieved here has been. What is interesting, however, is that the concert hall is an uncompromisingly modern building, about which I suspect many people will have strong views; like it or not, therefore, the legacy will not wholly be the sort of urban blandness that typifies much recent development in Britain in waterfront locations. A rather different, but none the less very valuable, further example of CMDC's role in these terms is the part it played in helping the British Council to move the majority of its home-based activities to Manchester, on a former gasworks site on one of the main southern approaches to the city centre.

2) CMDC has mounted a continuous programme of expenditure on environmental improvements. Particularly in the Castlefield and Rochdale Canal areas, but also in other parts of its area, CMDC carried on with and extended programmes that the City Council had itself been undertaking. There can be little doubt, particularly in the light of what has been said about the fate of the Urban Programme, that CMDC has been able to sustain an environmental improvement programme in its area sizeably bigger than the City Council itself would have been able to undertake. At the time of writing there is an issue as yet unresolved about how all the long-term maintenance of this will be funded, but this should not detract from what has been achieved here.
3) A significant boost has been given to the growth of city-centre housing. At the time of CMDC's emergence on the scene, there were clear signs that the corner was at last beginning to be turned in this policy area, but CMDC has undoubtedly given considerable further impetus to this momentum. A recent report about housing in its area (Central Manchester Development Corporation, 1995) shows that by the time that report went to press 2,290 housing units of various kinds had been completed at a total cost of just over £104 millions, with a further 135 units under construction adding nearly £7 millions to the total cost. At that time, 96 per cent of these units were either let or sold. CMDC's grant-aid contribution to this total expenditure package of £111 millions was just under £14 millions, or about 12.5 per cent, although of course this varied greatly between individual schemes. Of these 2,425 units, 688 (28 per cent) were recorded as being for sale, and 1,737 (72 per cent) were recorded as being for rent, although in the case of one such project either renting or equity share purchase were available as options. By far the largest single project in the renting category was a student housing project of 1,040 units, or virtually 60 per cent of the total rented stock recorded, which reinforces the point already made in Chapter 6 about the connections between the demand that the Higher

Education Precinct creates and the supply opportunity that the adjacent city centre's building stock provides. All this meant that by the mid-1990s the city centre was well on its way towards achieving the objective (shared by the City Council and CMDC) of having 10,000 residents by the millennium or shortly afterwards.

4) The quality of marketing and promotion of the city centre has undoubtedly been improved by CMDC's activities, and its willingness to participate actively in wider initiatives such as the Olympic bid has in turn been an asset to those promotions. CMDC recognised that in promoting its patch it had also to be promoting Manchester, and in so doing it undoubtedly added to the quality of this overall effort, which was not at a very high level before CMDC came along.

These are worthwhile achievements in their own right. Whatever view one takes about whether or not they would have materialised in the absence of CMDC is in a sense beside the point; from the city's perspective, they will be an enduring legacy. These achievements can be acknowledged for what they are without getting on to more contentious territory, which is CMDC's claim (p. 17 of the 1995 housing report, quoted above) that '. . . to date the Corporation has attracted £412 million of investment into the area which has brought many exciting developments and created a vibrant new area for workers, residents and tourists'. We will never know how much of this would have happened anyway, or how much of it is due solely to CMDC's efforts, but it is very hard to believe that nothing would have occurred at all since the area was not wholly dead before CMDC arrived. This is arguably more to do with public relations than it is a hard-headed assessment of the value that has been added through its efforts.

A critique of CMDC's period of existence would probably also include the following more negative points:

1) In its early years in particular, it was too prone to take the view that since it was there to encourage development most development proposals in principle should be supported. This had the effect of getting it into commitments in a couple of cases that would have benefited from more thought in the first place. It also resulted in the early years in a large accumulation of office consents, many of which (no doubt in part because of the subsequent impact of recession) did not get implemented. The effect of this was mitigated by CMDC's policy of granting most consents for a single year only, in order to keep the regeneration impetus moving, but it still resurrected in some quarters the issue of hope value which the City Centre Local Plan had been trying to dampen down.
2) It wasn't very active in the Piccadilly Station and the Pomona areas, which were the outer edges of its territory. Basically, CMDC divided its area into six subareas, and nearly all its activities were concentrated in the four sandwiched between these two areas. This was a pity in the sense that both these areas were potentially significant gateways to the city (Piccadilly

Station for obvious reasons, Pomona because it was the link in from Trafford Park and the area of Trafford Park Development Corporation), but they were rather longer-term regeneration opportunities than the other areas and thus received little attention from an avowedly short-life organisation.

3) Its conservation record was open to criticism, in the sense that its care for listed buildings and for developments in conservation areas none the less started from its regeneration remit.
4) It wasn't a democratic organisation, and made no claims to be. Thus, as compared with the principles of open government that applied to the way the City Council discharged its planning functions (meetings with published reports and speaking rights, for example), CMDC was an organisation that met in private with no public access to its papers or its decision-making processes unless for its own reasons CMDC chose to grant such access. One paradox of this was that, in encouraging (very successfully) population growth through housing development in its area, it was also denying to those same citizens the access rights to the planning system that they would have had elsewhere in the city.

The demise of CMDC on 31 March 1996 left a series of 're-entry' questions to be faced by the City Council, of which the main ones were probably as follows:

- How to maintain to an acceptable standard, through revenue budget arrangements that had not made provision for this burden, the sizeable increase in the amount of land CMDC had treated through its environmental improvement programme?
- How to maintain the momentum and the expectations in the area that CMDC's presence and actions had established?
- How to develop effective relationships with CMDC's customers, to the extent that these did not already exist?
- What to do about unfinished projects?
- How to address the problem of promoting the city, given that CMDC's promotional budget had played a major part in recent efforts of this kind?
- (for the planning service in particular): How to deal with the budgetary problem caused by the disappearance of development control agency fees from CMDC, given the removal from the budget for the planning service when this agency agreement was reached in the late 1980s of the mainstream resources for which this income was regarded as a substitute?

Overall, balancing achievements against criticisms, CMDC can in my view justly claim to have had a positive rather than a negative impact. The much bigger question than this localised balance sheet, however, is the extent to which all this amounts to the regeneration of the inner cities. The mini-UDCs, including Central Manchester Development Corporation, were a response to

Margaret Thatcher's statement on the steps of Conservative Central Office in the early hours of the morning following her 1987 general election victory: 'On Monday, we have a big job to do in some of those inner cities' (quoted in Robson, 1988, p. vii). Yet in Manchester's case, notwithstanding the propinquity of the city centre and the inner-city areas that surround it, in a sense a large part of the problem is that the interactions between these two are nothing like as strong as they ought to be if the problems of the one are to be ameliorated by the strengths of the other. In many ways, the city centre is an oasis in an inner-city desert, with what goes on in it in terms of its economy often having very little to do with the social and economic problems of inner-city residents. The work of CMDC can thus be seen as another example of the 'pepper-potting' of inner-city initiatives which has characterised top-down practice, whereby a small number of very focused high-profile activities are surrounded by much larger areas of decline and neglect, with the former not only having very little impact on the latter but also as a result making little difference to the general proposition that on the whole the problems of deprivation in the inner city are getting worse. This is not to say that in their own terms the achievements of CMDC were not worth having, and it is certainly not a criticism of the staff or of the board of CMDC, who could only get on with the job they were set up to do. But it is to question very seriously whether its efforts actually represent the sort of 'big job in the inner city' that Margaret Thatcher's statement might have led us to expect.

Policy approaches to the inner city from the bottom up

At the local level, approaches have tended to vary according to the nature of the problems of a particular area, resource availability of various kinds, and also the degree of activity and strength of feeling in local community groups. This latter point has had a particular resonance to it in terms of the Urban Programme, where for a time a significant share of the available resources was used to support community groups, voluntary organisations and similar bodies in a wide variety of ways. This was a response to what has been said above about the nature of inner-city problems; many of them are about the economic, social and environmental circumstances of people from day to day, and not all this is readily amenable to large-scale (or even small-scale) capital projects or best looked after by local government bureaucracies. The decline and then the disappearance of the Urban Programme caused real problems in these terms. The difficulty with this quickly turned out to be that it was much easier to give than it was to take away again. Community groups on the whole tended to be very pleased to receive revenue support to help them employ staff or provide particular services, but they then tended to regard that as a permanent arrangement. Taken to its logical conclusion, this would mean that in so far as Urban Programme resources went into revenue support for various organisations, that part of the budget was fixed and inflexible, as compared with a situation where capital resources became available again in future years once individual projects had been completed. These difficulties were compounded with regard to organisations that were not themselves direct service providers to immediate customers but which were indirect co-ordinating or resourcing bodies for others, because accountability in terms of what they were achieving was often hard to establish. These problems were resolved over time (although by no means to everyone's satisfaction) through a reduction in the share of the budget devoted to this type of support, and a balance within the resources available for these sorts of activities in favour of organisations directly delivering services known to be needed; but the process of withdrawing support from some organisations that this involved was politically difficult, and it was not helped by the argument that the council was reducing expenditure on support services itself as a result of mainstream revenue budget reductions. Some of the difficulties of these processes in the first few years of the Urban Programme are recorded from the perspectives of voluntary organisations in Williams (1983).

This downgrading of the significance of social schemes was partly a result of an increasing emphasis by government on economic and environmental projects as the main components of the Urban Programme. From a planning department perspective, this actually meant that our environmental programmes assumed greater prominence, and at any time during the height of this period the preparation pool for Urban Programme and Derelict Land Grant schemes was about £10 millions. These two in combination were a very valuable means of getting environmental improvement work done, for example in the east Manchester area where a great deal of industrial dereliction arising from the exodus of heavy industry needed to be tackled to improve the appearance and the general attractiveness of the area for its residents and potential investors and to treat some very degraded land. Often, the distinction between the Urban Programme and Derelict Land Grant was a matter of degree rather than of kind. There was certainly a financial interest from the City Council's perspective in having the Department of the Environment accept that a particular site was derelict, because this brought with it a 100 per cent grant regime rather than the 75 per cent grant regime of the Urban Programme.

In the east Manchester area alone, the City Council spent in excess of £10 millions of Derelict Land Grant resources between the mid-1980s and the early-1990s, helped in the later years by the acquisition of Rolling Programme status for east Manchester and the Irk Valley (on a joint basis with Tameside MBC, our neighbouring authority immediately to the east) which for a while gave greater continuity and certainty in terms of resource availability. My view is that this was an essential element in the process of turning east Manchester round from being the former 'workshop of the world' that had lost its *raison d'être* but retained much of the scarring that went with that former function to being an area where both housing and industrial/commercial investors were prepared to undertake projects again. This was seen as a role for the public sector; the long-term process of changing an area's image and environment and changing its land availability position was not something that the private sector had shown any interest in. Presumably this was at least in part because there was no prospect of a pay-off from all this in the short to medium term; this was long-term investment, which included some very badly polluted sites of which the private sector had been divesting itself over a considerable period of time.

The story of regeneration in east Manchester is told in greater detail in Tye and Williams (1994), but it is perhaps worth recording in the light of things that have been said elsewhere in this book that the planning framework that guided most of this work was an informal one, produced relatively quickly once it became clear that there was an opportunity to get something done. When it became clear in the early 1980s that there was a significant prospect of resources being made available on a scale which would enable at least some of east Manchester's problems to be tackled, the East Manchester Planning Framework was prepared. This was a broad-brush document that identified four key objectives for action, and then set out to provide a context for programmes in particular policy and geographical areas, whilst accepting that this would be an evolutionary process wherein detailed programme content would be determined in the light of prevailing circumstances. The four basic objectives were

- assisting the economic recovery of the area;
- improving the environment;
- improving the accessibility of the area and circulation within it; and
- enhancing the area's housing stock.

This was updated in the early 1990s once east Manchester had been chosen as the location for the main Olympic stadium and velodrome area by a regeneration strategy which emphasised the potential role of sports-led urban regeneration, but in practice apart from this emphasis the basic objectives were very similar to those declared a decade previously. These experiences certainly leave two queries in my mind, however, about the process of passing Derelict Land Grant to English Partnerships with its emphasis on the achievement of 'hard' end uses in the short to medium term:

- How is long-term commitment to be achieved to the process of upgrading the environment of a very blighted area to the point at which that area can once more become attractive to investors?
- How are the most degraded sites, where the need for reclamation may arguably be the greatest but where the prospect of 'hard' end uses in the short to medium term being achieved is significantly reduced by these very same forces, to get sufficiently far up the pecking order for funding to be made available for them?

It is interesting in this context to look at what the derelict land reclamation record in Manchester as a result of the high level of activity of the period from the 1980s to the early 1990s actually was. My report to the Environmental

Planning Committee of 17 November 1994 showed that ov...
1988/93 just over 110 hectares of land had been treated using Dere...
Grant resources. This was the equivalent of treating about 45 per cent o...
land accepted in 1988 as being derelict. What the report also showed, however, was that the amount of land classed as derelict at the start of this period was 249 hectares, and the amount of land classed as derelict at the end of this period was 326 hectares, or 30 per cent greater than the 1988 figure. In other words, the high level of land treatment activity over that period had still not managed to keep pace with the (mainly economic) processes that had been continually creating dereliction, although of course the locations and the types of sites changed as a result of the operation of the programme. Put another way, if for the purposes of simplicity it is assumed that all the land treated over the 1988/93 period had been classed as derelict in 1988, the amount of new derelict land created over the period (187 hectares, on this basis) was the equivalent of 75 per cent of the pool of derelict land that existed in 1988. This must reinforce the questions asked above about whether the sorts of contemporary regimes for handling derelict land will be sufficient in places like Manchester to tackle effectively what is clearly still a continuing and major problem.

It was also important to try to ensure that these city-wide environmental programmes were accompanied at the very local level by a facility for local people to be able to play a substantial part themselves in tackling a local environmental problem that they thought was of significance. As a consequence two particular local programmes of this kind were devised by the Planning Department, handled in terms of member-level decision-making through a small members panel operating with delegated powers, and funded through the Urban Programme. The first of these was the Community Initiatives Fund. This was a small-grants programme for local environmental groups to enable them to undertake a local environmental improvement scheme. The test was not whether the council happened to like the individual scheme, but rather whether it was a scheme from which the local community (as distinct from a small number of private individuals or groups) would clearly benefit and for which there was local support. Over the years of its operation, the Community Initiatives Fund spent well over £1 million of Urban Programme money on a wide range of community and local schools schemes, and was widely regarded when it ceased because of the demise of the Urban Programme as a very real loss to the people of the area. The other programme, which operated on a very similar basis, was the Disabled Access Grant scheme, which started off with an annual budget of £200,000 and increased from this. This was essentially about helping to make existing buildings which experience a volume of community or public use accessible to disabled people, and it was paralleled by a drive through the development control process to ensure that issues of disabled people's access were taken seriously. In combination, these two contributed to making Manchester a more accessible city for disabled people, although this process had not been completed when the programme in its original form stopped. For both these, the Community Technical Aid Centre was an important source of professional aid to grant-seeking groups, and as a consequence CTAC as an organisation received Urban Programme support via the council and the Planning Department, and a Planning Committee representative served on its management board.

Some of the most important economic questions to be addressed in the inner city relate to the need to try to ensure that the people of that area and particularly of its most deprived parts do benefit directly from programmes. As has already been said in Chapter 6, whilst a great deal is being done, and on the whole very successfully, to strengthen the city's economic base, this is in a sense at the macro level. There is always a worry about who actually benefits from major projects (Loftman and Nevin, 1995), despite very worthwhile things that can be done, for example to reach voluntary agreements through the Manchester Employment in Construction Charter (Manchester City Council and Jarvis Management Training, 1994) that new construction work both employs and trains significant proportions of local people. Just as doubts have already been expressed about trickle-down theories as part of national policies towards inner-city areas, however, so it must also be said that the same doubts exist about whether local actions always produce local benefits. The *prima facie* evidence is straightforward. Benefit dependency in the most deprived parts of the inner city remains stubbornly high; over 50 per cent of households in both North C and Central A (most of the areas in the ring immediately around the city centre; see Figure 7.2) were in receipt of housing benefit in March 1992, and the Manchester Community Care plan for the period 1995/98 says that '. . . more than a third of the City's population is now dependent on Income Support' (Manchester City Council, 1995a, p. 6). This relates closely to Will Hutton's 30:30:40 split of society (Hutton, 1996, pp. 105–10), where the bottom 30 per cent are absolutely disadvantaged and are dependent upon the state for benefits of various kinds. Of course, it can always be argued that these figures would have been substantially worse had it not been for all the local and city-scale economic work that has been done over the years, and I think there is probably some truth in this. None the less, there can be little doubt that the economic difficulties of inner-city residents are at the heart of many of the problems of the inner city, such as alienation, anti-social behaviour of all kinds and the creation of a 'black' economy linked in part to crime and drugs, as well as more obvious ones such as the ability of local shops and churches to survive; and as has already been said, there is evidence to suggest that these problems at least relatively are getting worse. There is a risk as a consequence that our cities will become increasingly polarised into the 'haves' (who don't live in the inner city, and in many cases work in the city but don't live in it at all) and the 'have nots' (who make up a significant proportion of the inner city, and indeed of other areas of the city as well), with all the potential consequences for social disruption that have already been seen. This distinction mirrors that between 'city needers' and 'city users' made by Gladstone (1976). The greater danger still is that if this becomes embedded as a way of life for people that is handed down from one generation to the next, it will become ever harder to tackle. I would say during my 16 years working in Manchester that there were clear signs that this was happening.

The obvious question that this gives rise to is why, despite all the things that have been done in the name of inner-city policy, does it appear in a place like Manchester that along at least some dimensions the problems are getting worse rather than better? This is an important question to address, because if we knew the answers to it we might well be a lot nearer to finding ways to improve what we are achieving. I certainly don't have a complete set of answers, but the following propositions may contribute to this debate:

- Local councils do not by themselves have the powers or the resources to tackle the root causes of social and economic deprivation. They get on with what they are able to do with the powers and the resources available to them, which often lag behind the expectations that are held of what councils can do. There are certainly proper questions to be asked about whether all this is done as well or as effectively as possible, but we should not try to pretend that local authorities can solve these problems through local action alone.
- There is a lot more that can be done through partnerships of various kinds with all sorts of agencies or interests, including directly with local people, which add value to what each can do individually. As yet, we are probably only at the dawn of our understanding of the potential of this instrument.
- There is no coherent and sustained national policy towards urban deprivation which focuses all the available and relevant national agencies and resources on how they can contribute effectively to the 'top down' components of the job to be done. Instead, we have had a plethora of government policy initiatives, which often amount in a very disjointed way to an ever more intensive chase by an ever-widening pool of competitors after an ever-reducing pool of money hemmed in by ever more restrictive sets of rules. The fact that it is often this or nothing tends to make local authorities feel that they have little choice other than to enter such competitions, although it could be argued that all this is very wasteful of the time and effort put in by those that compete, and especially by those that lose the competitions. What all this has tended to do, however, is to put the limelight on and the resources into individual initiatives, which are very worth while in themselves but cannot be an attempt to tackle inner-city problems in the round.
- Trickle-down economic thinking linked to property regeneration will not solve the inner city's problems, because at their heart the inner city's problems are not mainly about property regeneration. There is a perfectly respectable case in its own right for a focus on property regeneration, because the need for work of this kind should be part of the inner-city portfolio; but we should not delude ourselves into believing that this is doing much other than tackling property regeneration issues.

- We simply don't know enough about what works well and what doesn't work so well in the inner city and why, because there has been far too little research and hard-headed appraisal associated with nearly two decades of supposedly targeted policy. From this perspective the work of Robson *et al.* (1994) added very substantially to our understanding. It also demonstrated, in looking at Merseyside, the Manchester area and Tyneside, that there are very important differences within and between each of these localities. The theoretical arguments in favour of policy analysis and review are fairly well understood and accepted, but the practical politics of this are an entirely different proposition. There are many examples of ministers announcing their latest policy initiative amidst a fanfare of trumpets, but far fewer of ministers announcing subsequently that it has been demonstrated that some aspects of their policy initiative (or that of their predecessor, since there have often been several changes in the interim) have not worked. Indeed, the strength of the political commitment to a policy initiative from the outset can be a major barrier to effective evaluation of what that policy is achieving. Equally, quangos set up to do a particular job usually produce glossy public relations reports saying how much they have achieved, which are not often models of reflection or of self-criticism. At local level too the political investment put into policy initiatives tends to mean that they have to be regarded as successes come what may. The consequence of all this is that we are not learning from our experiences as much as we should, and we are probably repeating many of our mistakes. An interesting recent report which does begin to analyse what innovative approaches may be able to achieve in terms of the generation of urban employment is by Dalgleish, Lawless and Vigar (1994), but there is far too little of this kind of material available.

One of the major conclusions that the City Council has reached over the years in trying to grapple with some of these issues is that there is a need to try to reverse the historic trends in the inner city of depopulation and social polarisation. The economic, social and demographic forces that have shaped the present situation, as this chapter has tried to demonstrate, are extremely strong, and so this objective will not prove easy to achieve. At the same time, the benefits that would be experienced by the inner city in terms of a sounder economic base and perhaps a more 'balanced' set of communities (an elusive concept that means many different things to many people) are potentially very worth while; and at the very least, it is an objective which wouldn't if it could be achieved condemn the inner city to a state of permanent hospitalisation, which it could be argued would be the inevitable outcome of a continuation of recent trends.

The City Pride process

When the Secretary of State for the Environment launched the City Pride initiative late in 1993, it amounted to an invitation to the cities of Birmingham,

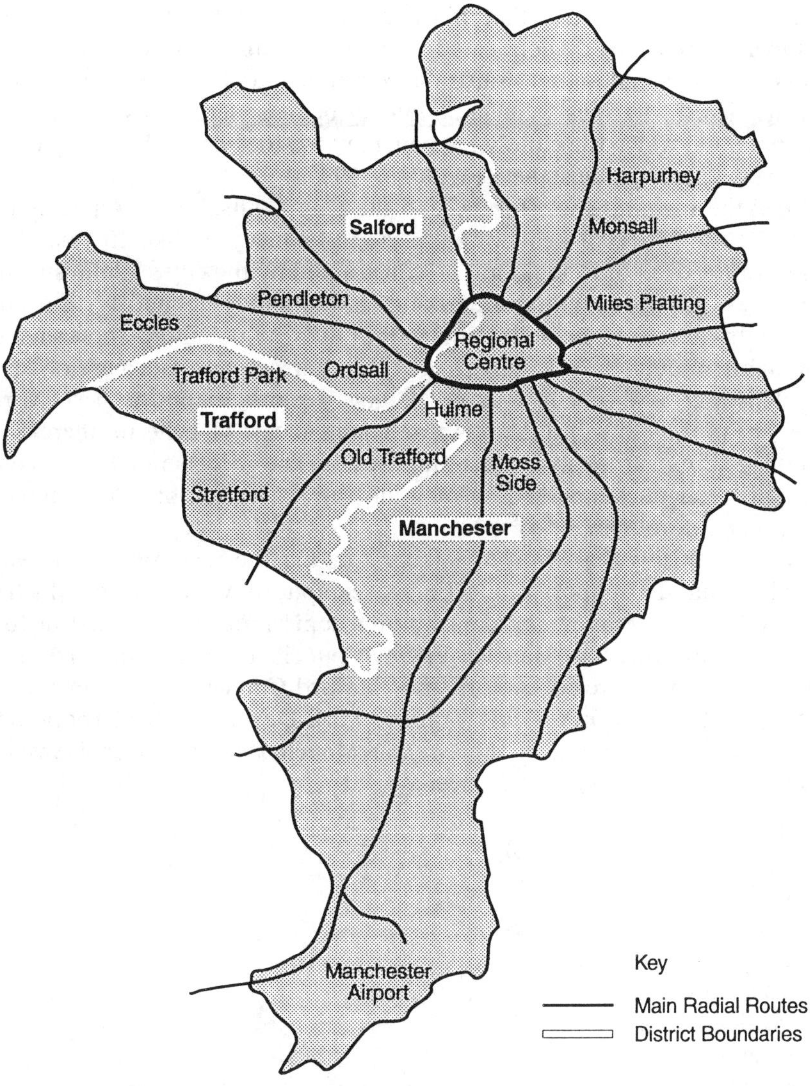

Figure 7.4 The Manchester City Pride area

London and Manchester to produce a prospectus setting out at least a medium-term vision of what was needed to regenerate themselves successfully, in harnessing through a co-operative process as many interests as possible which had the welfare of their cities at their hearts. It was clear that there was no specific money earmarked for this initiative, but it was equally clear that the government had no real idea of what sort of Pandora's Box it might be opening in issuing these invitations. It was said at the time that one of the formative influences in the government's thinking about this was the experience of the large amount of support the Manchester Olympic bid had generated, and in particular that ministers were very taken by the value that the Olympic bid had demonstrated of a banner under which the various sectors in a city could unite towards the common end of promoting that city's interests. The invitations were undoubtedly open ended; this was a tool of some potential which the cities were being invited to shape as best they could in their own local circumstances (Williams, 1995a).

Whilst the invitation was formally to the City Council, it was quickly decided that the process of covering a period of something like 10 years in a prospectus of this kind should not be constrained by the city's administrative boundaries. Accordingly, the area that was ultimately defined for the purposes of this exercise included a large part of the City of Salford and a smaller (although still quite substantial) part of the Metropolitan Borough of Trafford. This area is shown in Figure 7.4. This westerly extension perhaps reflected the point that whilst there is a well defined east Manchester there is no such thing administratively as west Manchester. The chosen area thus encompassed all or significant parts of the administrative areas of three local authorities and two development corporations.

The mechanisms that evolved for undertaking the necessary work were tripartite in nature, supported by officer-level inputs of various kinds. Much of the detailed work was undertaken by nine topic groups, convened in four cases by senior officers of Manchester City Council and in the other five cases by the Chief Executives of Salford and Trafford Councils, the Chief Executives of Central Manchester and Trafford Park Development Corporations, and the Director-General of the Greater Manchester Passenger Transport Executive. The nine topic groups were

- industry and commerce;
- housing and energy;
- crime;
- health and personal services;
- arts, sports and culture;
- transportation;
- education and training;
- marketing; and
- regional centre.

Initially, invitations to membership of these groups arose from a brainstorming session by Manchester City Council officers (including myself) which sought to generate appropriate cross-sectoral membership whilst keeping the numbers of people down to a level at which a working party could actually work, but in practice topic groups then augmented their memberships as they saw fit. City Planning Department staff contributed throughout to the work of the regional centre and the transportation topic groups, and to other groups on a less continuous basis when requested to do so. The work of the topic groups was integrated and co-ordinated by an executive team, chaired by Howard Bernstein (Deputy Chief Executive of Manchester City Council) and consisting of the convenors of the topic groups and a small number of other Manchester City Council staff including myself. This in turn reported to an advisory panel, chaired by the Leader of Manchester City Council and acting as a forum for discussion of the strategic direction of the exercise. This was a loose meeting consisting potentially of nearly 90 members drawn from the various organisations participating in the process; indeed, the total number of participating organisations including those represented on the various topic groups was about 150.

Each topic group was asked to produce a statement of strategic intent and a series of key projects or ideas that would move towards the achievement of that intent, in response to what was ultimately expressed as the City Pride vision and as seven key themes. These had been the subject of some initial discussion early in the process, and they were refined and re-expressed as more people and organisations got involved; but in truth there wasn't a vast amount of debate about them nor obvious signs of disagreement with them. The vision and key themes are set out in Box 7.3.

The form of the document that was eventually produced (Manchester City Council, 1994c) follows this vision and these themes through various topic chapters to a series of project proposals, via a high-standard, 'glossy' brochure-type publication which is quite brief (70 pages) given the span of its coverage and the amount of space devoted to various kinds of illustrations. When it was launched in September 1994, it was the first of the three City Pride prospectuses to emerge; and indeed, it was still the only submitted document by the time the government responded early in December 1994, although various drafts had been circulating from the parallel work in Birmingham and London. In its own terms, the response was very encouraging. Its essence (letter from John Gummer, Secretary of State for the Environment, to Graham Stringer, Leader of Manchester City Council, dated 8 December 1994) was as follows: 'In short, I am encouraging you to prepare an Action Plan to take the proposals forward, not least to ensure that these strategic projects and proposals complement and enhance mainstream funding and investment in the area, including Single Regeneration Budget and European Union funding.'

As had been known from the outset, this was not a process which had any funding directly attached to it, and thus it was neither a surprise nor a disappointment when the response contained nothing specific in these terms. At the time of writing, however, the proof of this pudding remains in the eating. The process of action planning referred to above is underway, and individual projects are indeed being taken forward in various ways, although it could not really be said that there is yet much by way of an outcome if this is measured in terms of specific resource commitments that might not otherwise have been obtained. In terms of the operation on a continuing basis of the topic groups, this is already causing problems of loss of momentum in some cases, because it isn't clear in this kind of situation what the focus of the work of the topic

Box 7.3 Manchester's City Pride statement: vision and key themes

The Vision Statement reads as follows:

'Manchester and neighbouring areas of Salford and Trafford by the year 2005 will be unchallenged as:

- a European regional capital – a centre for investment growth not regional aid;
- an international City of outstanding commercial, cultural and creative potential;
- an area distinguished by the quality of life and sense of well-being enjoyed by its residents.

This vision rests on the marrying of an enhanced international prestige with local quality and benefit. Neither can be achieved without the other.'

The seven key themes were as follows:

- The repopulation of the city, through policies encouraging a higher density and a strengthening of physical and social fabrics.
- A reduction in the dependence of Manchester on the northwest region through the attraction and development of multinational companies and international institutions, notably from the European Union.
- An emphasis upon impoverished and disadvantaged sections of the community benefiting fully from developments in order to reduce unemployment and poverty.
- Buildings and facilities must be accessible (in all its senses) to all sections of the community.
- The city's infrastructure needs to be brought up to the standards achieved in many European capitals, for the benefit of residents, visitors and investors.
- Concern for the environment and for sustainable development, so that quality of life in the city improves in terms of the balance to be struck between economic, environmental and social considerations.
- The voice of Manchester needs to be strengthened to enable it to compete effectively with the developing city regions within the European Union.

groups now is. On the other hand, the value of the process in some senses has already been achieved; a broad prospectus now exists which has a very wide range of support and thus represents a strategic starting point for many other things, and the process of intersectoral co-operation has been further developed. Indeed, in all probability it will be the City Pride process rather than its specific product that will be the enduring legacy here. Whilst this should not be confused with widespread public and community consultation and involvement, because it was really the *cognoscenti* talking to each other, this was in itself both a desirable thing and a substantial gain; and it built upon the growing tradition in the Manchester area of intersectoral co-operation that had already marked the Olympic bid, and which created improved networks of people willing to work to common ends and with each other through the informal contacts they had built up.

The City Challenge process: the regeneration of Hulme and its wider manifestations

When Michael Heseltine announced the City Challenge initiative in 1991, Hulme had already been established by the media and by politicians as the quintessential failed inner-city estate of the 1960s/1970s redevelopment phase. In particular, it reflected the failure of large-scale prefabricated housing technology, embarked upon as a means of increasing the productivity of the British housebuilding industry to meet predetermined political housing targets but at the expense of the people who were destined to occupy the end-product. It also reflected a failure of layout, having turned Hulme from a grid-iron of densely packed terraced streets very well connected into the street network of the rest of the city to an inward-looking estate with a complex set of loop roads very poorly connected into the rest of the city. Many of the publicity stills released by the Department of the Environment at the time of the launch of City Challenge and subsequently appeared to show Michael Heseltine standing in front of one of the Hulme crescents, as if this typified not merely the problem but also the arrival of the knight in shining armour who was going to solve it.

The fundamental problem in Hulme had been the prefabricated deck-access blocks. Their inability to withstand the Manchester weather (alongside many other physical defects) had quickly rendered them both unsuitable and unpopular as the kind of family housing units that they had been designed to be. These weren't the only problems with Hulme in its redeveloped form. The layout difficulty has been referred to already above; as a creature of its times, the layout of Hulme was trying to achieve in its own way a pedestrian-friendly environment with a high level of segregation from traffic (including a pattern of upper-level walkways linking housing blocks) without being aware of the difficulties that some of these measures would in turn bring. In addition, the area was full of open space (it looked very green from an aerial photograph), but most of that open space was informal and incidental, with little by way of a clearly thought-out function and over the years a very poor aesthetic quality because of inadequate maintenance regimes. The sheer ugliness of in particular the deck-access crescents (ironically, named after people with Bath connections, according to some stories because the Bath crescents had been the inspiration for the layout of Hulme) also made the whole place visually a very depressing experience, which promoted in turn a lack of care for its physical environment by residents and visitors. A long list of economic and social problems could be added to this series of physical manifestations, together with the problems of stigmatisation that an estate of this kind almost inevitably experiences. The consequence of all of this was that the City Council during the first part of the 1980s was decanting family groups with children from the estate or into low-rise homes elsewhere in Hulme, and increasingly the flats and maisonettes regarded as habitable were let to single or groups of young people, many of whom were students at institutions in the adjacent Higher Education Precinct. This created a very distinctive, and in some ways very vibrant, community life in Hulme, but arguably it also increased the area's sense of isolation from much of the rest of the city, precisely because this was increasingly an 'atypical' community. From the council's perspective, this was in many ways a stop-gap measure designed to get at least some use (and some income) out of what was increasingly seen as a badly failed redevelopment until such time as radical solutions could be found which were capable of being implemented. City Challenge provided that opportunity, because it introduced the possibility of a resource level that could contribute to a comprehensive approach to the problems of the area.

Although City Challenge was a competitive bidding process, it would have been unthinkable if it was genuinely intended to focus on the greatest problem areas in Britain's inner cities for it not to have led to the selection of Hulme as one of the winners in the first round. Whilst the City Council could not take success in these terms for granted, there was undoubtedly a high level of expectation in Manchester that this would be the outcome; and so it proved. One of the component elements of this was a very strong belief that history should not repeat itself. So, not only was the new Hulme to be reconstructed in such a manner as to link it back into the city (and it had major locational advantages in these terms, being immediately south of the city centre, immediately west of the Higher Education Precinct, and athwart the major road from the airport to the city centre) but also the sort of environment to be created was one which emphasised typical streets and a relatively low-rise albeit still fairly dense built form. Much of this came as a result of an intensive and an extensive series of consultations with Hulme tenants, because one of the other lessons that had been learned from the previous experience of redeveloping Hulme was the difficulties that can arise when for all practical purposes there is no real involvement in the redevelopment of the area by the people who will in future be living in it. Planning Department staff played a large part in this process before Hulme Regeneration Ltd (the new organisation created as a result of winning City Challenge to manage and oversee the regeneration process) came on the scene, and got very actively involved in much of the lobbying that was going on at that time. Hulme tenants' representatives also took an extremely robust view about how these processes should be handled, involving a high level of politicisation of issues and the involvement of some extremely articulate and intelligent people. I had the experience, for example, of speaking in London at a conference on environmental education, and of finding that I was harangued about the perceived failings of Manchester City Council by a Hulme tenant. She had come to the conference specifically to raise Hulme points in public, not because I was on the list of speakers as much as because a government minister was; and this was not a unique experience. What it meant, however, was that there was a high level of understanding about the aspirations being expressed on behalf of the people of Hulme and a desire on the City Council's part to ensure that these views were an essential component in taking the process forward.

As well as the necessary involvement of Hulme tenants, the council also needed to have a private sector partner if the terms of City Challenge were to be fully complied with, and this function from fairly early days was performed by AMEC (a northwest-based construction company with a strong regeneration arm which had worked with the City Council on other major projects). The board of Hulme Regeneration Ltd was established from these three elements: the City Council which had three board members drawn from the political leadership, AMEC and Hulme tenants. To help to get the new organisation running quickly, the Planning Department contributed two of its key members of staff who had been working for a high (and increasing) proportion of their time on Hulme anyway as secondees for the anticipated five-year life of the organisation. It had also undertaken as a basic resource document a detailed study of the social, economic and physical circumstances of Hulme. Internally, the City Council reconstructed itself by establishing a Hulme Subcommittee of the Policy and Resources Committee chaired by the Leader of the Council to operate with delegated powers as a focus of decision-making on Hulme, including development control; and the council's Deputy Chief Executive assumed a co-ordinating function in order to support the work of the subcommittee, although the Planning Department took planning application reports to it in the usual way.

The Planning Department had already provided some basic strategic material on the redevelopment of Hulme in the light of the consultations and discussions described above, as a contribution to both the City Challenge process and to the draft UDP. One of the early tasks of Hulme Regeneration Ltd was to follow this up in late 1992 and in the early part of 1993 so that a more detailed idea of what the redevelopment process was trying to achieve could be established, and in particular so that some early projects could be proceeded with both to begin spending City Challenge resources and to make a clear statement of intent. One of the mechanisms used was a community planning weekend, in which my participation is recorded photographically in the *Hulme Development Guide* (see below). These early experiences quickly led to the view that what was needed was a guide to developers and to other interested parties setting down the requirements and the aspirations of the new Hulme. The decision that such a thing should be generated turned out to be much more straightforward than the process of producing it, however, and it wasn't until June 1994 that an agreed version could be published (Hulme Regeneration Ltd, 1994) after several previous drafts had been rejected at one level in the process or another.

The basic principles that governed the *Hulme Development Guide* were as follows:

- The clear definition of the public realm, defined as all the outdoor spaces between buildings where people interact, as a priority in environmental design.
- An emphasis on streets and squares, including in particular the re-creation of Stretford Road as a high street for Hulme.
- A rich mix of uses and tenure patterns in Hulme. The mixed use doctrine in all its forms was seen as being essential to the successful redevelopment of the area.
- A high enough density to ensure that Hulme has a sustainable population and economy.
- Strong links both to and through Hulme to create high levels of 'permeability'.
- A strong sense of place, fostered by the location of major buildings at nodes of activity.
- All spaces should have clear functions which promote stewardship of their environment on the part of Hulme's residents.
- Urban sustainability in all its forms (but not very clearly defined).

These principles then found expression both in a series of design statements meant to offer guidance as to how they should be translated into practice and through some technical recommendations as to appropriate standards of various types. It was clear that this was intended as a guide and not a template, although it also became clear that different people interpreted these propositions in different ways; from the zealots who regarded it as being tablets of stone, to those who thought they could ignore it whenever they wanted.

One of the driving principles behind the *Hulme Development Guide* was the view that it took about the relationship between the design of the built environment and crime. This was a major issue in Hulme because its crime levels were high and (at least as important) because the fear of crime was widespread. The guide espouses the philosophy that the most effective way to reduce crime levels is to have lively and active streets and spaces, naturally overlooked at all times by people going about their normal business, and with a density of development and a mix of activity that enabled these things to happen in unforced ways. It turns its back on a view, in many ways espoused by the Greater Manchester Police through its Architectural Liaison Service, that tries to approach the problem of crime through making the target of crime difficult to enter and seeks to minimise from practical policing experience possible escape routes and hiding places. In no sense is this merely a philosophical difference; it makes a major difference in practice to the kind of built environment that gets produced. An illustration of this is given by the vastly different views taken to the low-density residential *cul-de-sac* typical of many suburban layouts, where the number of escape routes can be minimised and within which the individual house is made as secure as possible, to the point at which in recent times this approach has become a marketing advantage claimed by some private housebuilders. This view is anathema to the *Hulme Guide*, which sees it as producing the opposite of the kind of area it wants to create and as bringing along in its wake many other problems such as an inadequate density of people to support local community services. It is an approach which fits much more comfortably with the philosophy of 'target-hardening' and the concept of 'secured by design' associated in some quarters with police advice on these matters. Extensive debates about these kinds of issues, with limited evidence that the two approaches could be brought together, was one of the reasons why the guide took so long to produce.

The *Hulme Development Guide* was seen both politically and by its authors as having a much wider significance than merely the regeneration of Hulme, as is evidenced by its title: *Rebuilding the City: Guide to Development in Hulme*. This caused the Planning Department to receive some public criticism from the council's political leadership during January 1994, several months before the document was published, to the effect that the department was not carrying out council policy by applying the *Hulme Guide* to the rest of the city. The response that this was difficult to do in advance of the document's existence, and in particular in advance of having an agreed document in any form that staff could work to with any consistency, or could make available to private developers, did not go down well because it was said that we ought already to know what was wanted. Furthermore, it was argued that it was our past failings that had demonstrated the necessity for a policy stance which sought to apply the guide across the city as a whole, echoing the point made in Chapter 2 about the planning service continuing to bear the brunt of some of the criticisms of the failures of the inner-city redevelopment process some 20 years and more earlier. The basic operational problem faced by the planning service at that time (the latter part of 1993 and the early part of 1994) was that there were several different versions of what exactly was being attempted in Hulme in the heads of several different key players, and very little thought had actually been given to the questions that would have to be addressed in transplanting this once it had been agreed to the very different circumstances that would arise elsewhere in the city. All this was very difficult territory for the planning service, especially since the status of planning policies had been clarified in the early 1990s by emphasising through Section 54A of the Planning Act the primary role of the development plan in development control decision-making. As the UDP timetable described in Chapter 4 has already shown, this debate about the *Hulme Guide* and its wider application post-dated the UDP deposit draft public inquiry which was held in the second half of 1993, although we had been able through the inquiry's mechanisms to add into the UDP at Hulme Regeneration Ltd's request some useful extra policy material about Hulme itself.

The process of producing a guide based upon Hulme principles that covered the city as a whole did not bear public fruit until May 1995, however, when a draft city development guide (Manchester City Council, 1995c) was produced as a basis for public consultation. This guide was the work of an independent advisory panel whose members were drawn from a wide range of professional and academic backgrounds, in turn advised by a group of City Council officers including a senior member of the Planning Department's staff. The Leader of the Council, in a foreword to the draft, described its broad purposes as follows:

> Manchester's regeneration strategies are rooted in the need to create a more attractive and a safer City, attractive to both residents and businesses. We can best do this through the integration of economic and social activities; the introduction of high density development and diverse tenures; the pursuit of design principles which create safer areas and streets; the construction of streets, squares and buildings of variety and quality; and the development of strong vibrant neighbourhoods which can become a cohesive and sustainable part of the City.

The draft city development guide was thus aiming very high, and was attaching a great deal of importance to the form of physical layout achieved. A detailed textual comparison between that document and the *Hulme Guide* would show unambiguously the strength of the grounding of the one in the other. At the same time, the problems of transplanting advice from one relatively small and very special major area of regeneration to a whole city as diverse as is Manchester are enormous. Listing six of the major differences that would have to be overcome in making this jump should be sufficient to establish this point:

1) The city contains a wide variety of place types that bear very little relationship to Hulme. Examples include the city centre, the city's conservation areas and its areas of leafy suburbs, particularly on the south side.
2) In a polyglot city, a much wider range of public choices could be expected to be expressed than could be accommodated in one relatively small major regeneration area. For example, in a democratic society if people choose to live in suburban residential *culs-de-sac*, is it really being said that in that case they should go and live somewhere else?
3) Most development in the city does not come along in the form of major redevelopment chunks, as in Hulme, but is on small sites with a significant infill element in relation to surrounding development.
4) The controls available to secure conformity with a guide are much wider in Hulme than they are in most parts of the city. For example, the City Council owns most of the land in Hulme, and is able through the grant regimes associated with City Challenge to provide financial support for appropriate and desirable schemes. Neither of these conditions hold true in most of the rest of the city, where reliance would have to be placed very largely on the much more limited powers of the development control process.
5) The macro-objectives that govern development in Hulme are very different from those that apply elsewhere. Hulme is about achieving

regeneration within a short period of time in a clearly defined space, whereas managing change across the city as a whole is both continuous and timeless.

6) The difficulty in generating a public consensus is inevitably much greater across a whole city as compared with inside one redevelopment area, where there was already a considerable level of support for the essence of what was being attempted.

It was not wrong to attempt to produce a development guide for the city as a whole based upon Hulme experience, but it raised very major questions which ought to have been thought out carefully before embarking upon an enterprise of this nature. There was little effective discussion of these issues, however, because the council's political leadership had already decided that the Hulme approach was so impressive that it had to be applied much more widely. Comments were therefore sought only on the contents of the draft document and not on how these could be applied. Attempts to raise questions about the implications of the issues itemised above, which I certainly regarded as being legitimate matters which needed to be properly discussed, were treated by the council's political leadership as showing lack of commitment to what was being done which almost amounted to disloyalty, and resulted in stories appearing in the local and the professional press about policy conflicts. To my mind, this was largely an artificial row designed to stifle debate about the validity of political preconceptions that ought to have been tested. This was bad enough in itself, but what was much worse was that it gave the impression that the planning service was seeking to obstruct desirable change (which it wasn't), and also left in some people's minds the suggestion that the service was opposed in some ways to the work of Hulme Regeneration Ltd (which it certainly wasn't). Instead of the needed debate, therefore, a political process of manufactured conflict emerged, which was to rumble on across a range of issues related to the regeneration of Hulme.

Problems with the political leadership: the Stretford Road case

The difficulties referred to above continued, and in some senses were exaggerated still further, in what turned out to be a very public row about a development control case which highlighted Hulme Regeneration Ltd's aspirations to restore Stretford Road as a major east–west spine road for the new Hulme. This case study is drawn wholely from published council reports and from reports in the *Manchester Evening News*, which gave the case extensive coverage.

The Stretford Road case was actually about a planning application by Manchester Metropolitan University to build quite a large block of student residences on a piece of land that the university owned at the interface of Hulme and the Higher Education Precinct, although in planning policy terms it was regarded as part of the precinct rather than part of Hulme. It became known as the Stretford Road case because the effect of constructing the student

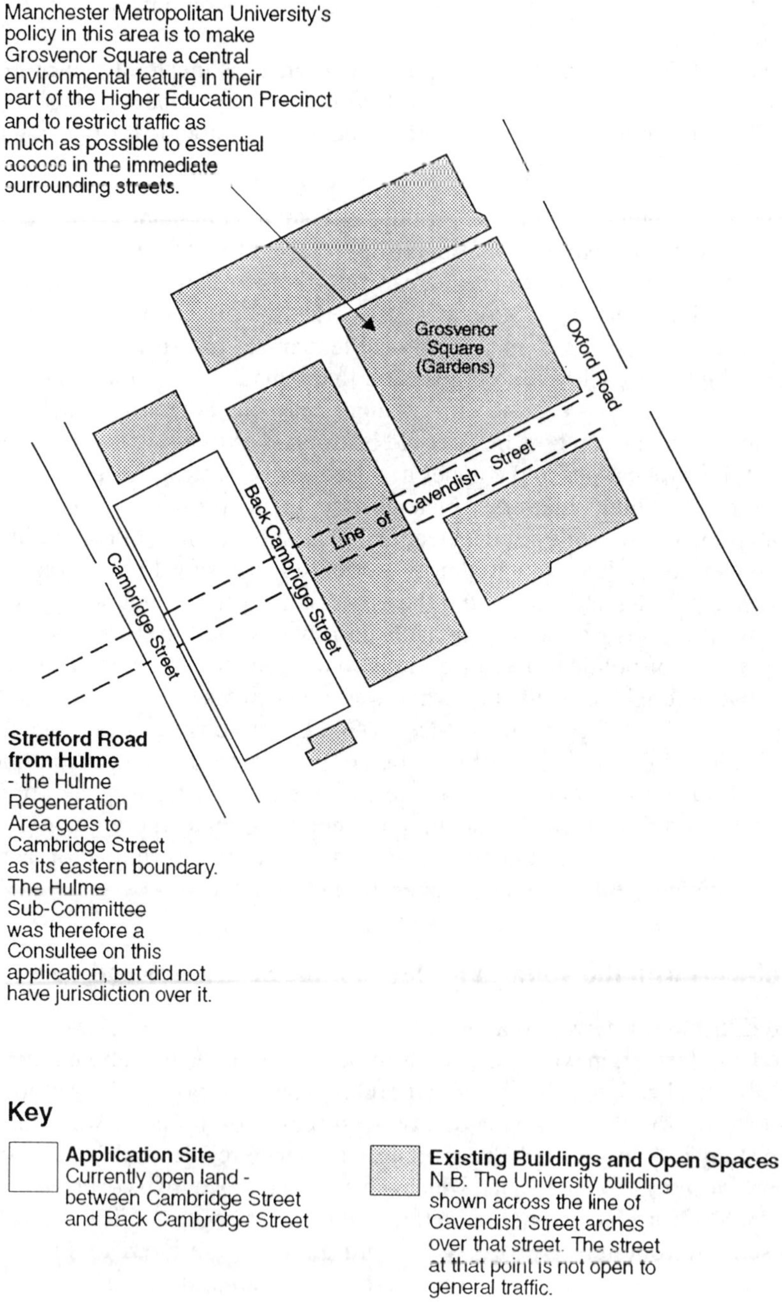
Manchester Metropolitan University's policy in this area is to make Grosvenor Square a central environmental feature in their part of the Higher Education Precinct and to restrict traffic as much as possible to essential access in the immediate surrounding streets.
Grosvenor Square (Gardens)
Oxford Road
Line of Cavendish Street
Back Cambridge Street
Cambridge Street
Stretford Road from Hulme
- the Hulme Regeneration Area goes to Cambridge Street as its eastern boundary. The Hulme Sub-Committee was therefore a Consultee on this application, but did not have jurisdiction over it.
Key
Application Site
Currently open land - between Cambridge Street and Back Cambridge Street
Existing Buildings and Open Spaces
N.B. The University building shown across the line of Cavendish Street arches over that street. The street at that point is not open to general traffic.

Figure 7.5 The Stretford Road case

residences would be to prevent the extension of Stretford Road as the east–west spine of Hulme as a vehicular route through this part of the precinct to Oxford Road along the line of the former Cavendish Street (see Figure 7.5), although in its final form it would not have prevented pedestrian and cycle access. Stretford Road had long since ceased to operate as a through route in this part of the city, although it had been part of the earlier grid-iron layout of this part of the city. In the intervening years, Manchester Metropolitan University had built an academic block across the former alignment with what was in effect a tunnel through it, thereby narrowing the size of the available route anyway.

The University's plans for the area anticipated major expenditure on the creation of what would be the central space of that part of the precinct, and they saw the idea of a through road in this area as being totally incompatible with their plans. The draft Manchester UDP supported what the university wanted to do here in terms of student residences and of environmental improvements, and both made no mention of the extension of Stretford Road through to Oxford Road. Whilst the submitted City Challenge documentation for Hulme talked about recreating Stretford Road as a spine for Hulme, it did not specifically talk about taking it as far as Oxford Road, which in any event was outside the City Challenge area. Thus, there was no written policy basis for the concept. The idea that the university's proposal should be opposed in order to retain the policy option of extending Stretford Road emerged from discussions within Hulme Regeneration Ltd as the process of dealing with the application unfolded; the City Council had three members on the board of Hulme Regeneration Ltd, including the Leader of the Council. It went from there to the Hulme Subcommittee, which was chaired by the Leader of the Council, where it was endorsed as a policy stance despite the fact that the land in question was actually outside its defined area of responsibility. At the time it was determined, this stance was never put to the council because the Hulme Subcommittee operated with delegated powers. The planning application fell to be determined by the Development Control Subcommittee of the Environmental Planning Committee because the site in question wasn't in Hulme, and the Hulme Subcommittee's role was that of consultee. This created a straightforward clash of views, because as noted above the formal planning policy position was both clear and straightforward, and therefore my advice to the Development Control Subcommittee was based upon that position.

This case was around in various forms from the end of 1993 to the middle of 1995. During part of this period in 1994, some aspects of the scheme were being reconsidered by Manchester Metropolitan University as a result of adverse planning comments, which were not about these broad policy issues but about the detail of the layout and in particular about the difficulties its original form would have caused for pedestrian and cycle movements. The momentum began to build in late 1994 when the university came back with a modified proposal which took account of these matters, which in effect meant that apart from any views that might be expressed about the design of the buildings themselves the only outstanding issue was the policy stance towards Stretford Road that was going to be adopted. The local press began to take more interest in the case.

Who exactly did what in this situation would be difficult to establish without the journalists involved revealing their sources, which of course they won't do. What began to happen, however, was that internal documents exploring these issues in fairly robust terms were made available to the press, and I found myself reading about the views of various parties (including myself) when I had thought that this was an internal process of sorting ourselves out. I have no doubt that what was going on here was an attempt at news manipulation; people were trying to get stories placed in the local press by releasing information and documentation in order to improve their chances of what they regarded as a successful outcome. Whether this is regarded as leaking or briefing I suppose depends upon your point of view, but I knew that I wasn't doing it and I also knew that the accumulating story was in public relations terms very unfortunate for the council. It also served to alert ordinary Labour members who were not directly involved in this issue to the strength of feeling in the political leadership about the Stretford Road argument. It is important in understanding this case to stress the point that the strength of feeling was coming from a very small group of people. The City Council only had three directors on the board of Hulme Regeneration Ltd (the Leader of the Council, the Deputy Leader, and the then Chair of the Housing Committee, all of whom also served on the Hulme Subcommittee), but they were each in powerful positions in their own right and collectively they constituted a large part of what most people meant when they talked about the political leadership. This process of political power was used first to ensure that the Development Control Subcommittee did not use its delegated powers on this matter, thus ensuring that it went to the City Council (and therefore to the whole Labour group) for a decision, and then secondly to secure support for the leadership's view that this application should be refused when the matter went to the full Labour group. This therefore ignored the planning advice, both about the policy position and about the high probability that a refusal would generate an adverse costs award at appeal because of this absence of a planning policy basis for such action. The City Council of 8 March 1995 decided that it was minded to refuse the application for the following reasons:

> a) for the reasons set out in the letter of 7th February 1995 from Hulme Regeneration Ltd. the design is unsatisfactory;
> b) the application is inconsistent with the proposals for the regeneration of Hulme and the Hulme Design Code; and
> c) Council wishes at this stage to protect the line of Stretford Road for vehicular traffic through to Oxford Road as far as possible in order to support the aspiration of the Hulme Sub-Committee and Hulme Regeneration Ltd. pending the receipt of detailed reports from the City Planning Officer and the City Engineer and Surveyor to the Environmental Planning and Highways and Cleansing Committees as a matter of urgency on the procurement and implementation of such a scheme;

> and the Chief Executive be instructed to hold urgent discussions with Manchester Metropolitan University regarding their application with a view to agreeing changes to the scheme to deal with the design problems and to safeguard the line of Stretford Road without significantly affecting the scheme.

The first two of the stated reasons for refusal were simply make weights; the letter from Hulme Regeneration Ltd was merely trying to put in more technical terms in relation to this application the views already reached by its board, and the arguments about inconsistency with the Hulme proposals would have been difficult to justify in terms of the written record. Nobody was arguing that by themselves they could possibly stand in the circumstances of this case as reasons for refusal. The meat was in the third reason, and the council's resolution in these terms more or less gave the game away. In all probability, however, the council could have toughed this out in the local press (although almost certainly not at appeal), were it not for the fact that the Leader of the Council in parallel with this process had written to the Chief Executive in effect asking him to investigate and report back on why the policy circumstances were as they were and not as the Hulme Subcommittee wanted them to be. This material too was subsequently leaked to the press, with both the Chief Executive's analysis and my strong response being given splash treatment.

This isn't the place to go into the content of this documentation; the point here is that the council's internal differences were being leaked by several parties (I know this to be true from direct feedback from press sources) presumably in order to seek to use the coverage in the local press that this would generate to influence events in their particular direction, and without apparently worrying too much about the fact that for the council as a whole this was the very opposite of good public relations. What appeared to matter was who won, and not how much damage was caused in the process. It was also, of course, grist to the mill for a wider article and a follow-up editorial about the style of leadership in the council in terms of the use of the 'public pillory' for chief officers whose advice was unwelcome which appeared in the *Manchester Evening News* of 25 and 26 April 1995; this was one of several cases cited in that article of how chief officers were treated.

The denouement of the case itself came fairly quickly afterwards. The process of negotiation between the Chief Executive and Manchester Metropolitan University, to which I was not a party, produced absolutely no willingness on the university's part to change its basic stance on the question of Stretford Road. There was very little surprise to me in this, because I presume that the university's professional advice about the planning circumstances of the case must have been fairly similar to the advice I had given the council, and it must therefore have known that its chances of winning at appeal with costs (which it had already said publicly it would be seeking) were very good. As a consequence, it was never clear to me how the centre of the authority thought it was going to persuade the university to depart from what was very likely to be a winning position. Thus it was no surprise when it couldn't, and as a result some essentially cosmetic changes were agreed to the design in order to be able to say that the council had got something out of the process; and the council at its July 1995 meeting backed down. Interestingly, the press coverage of this as a news item was much less strident than had been the coverage of the internal difficulties that had been laid bare as part of the process.

To my mind, the planning issues here were relatively straightforward. Under the British planning system as it is at present, a developer is entitled to the benefit of the provisions of the development plan as it stands at that point in time, and councils cannot unilaterally decide that they want to do something else and then argue that such an aspiration (because that is what the extension of Stretford Road to Oxford Road was described as being) overrides the provisions of the development plan. This was the case here, and this was the proposition on which my planning advice was based. What was difficult about this situation was that a small number of members of the political leadership did want to do something different without being prepared to accept that the current planning situation precluded this, and certainly without securing the agreement of the party which would have been most affected by that action (Manchester Metropolitan University). Nobody at the centre of the authority seemed able or willing to tell the political leadership that what they wanted in these terms was not deliverable, but instead a process of wish fulfilment was embarked upon. This ultimately foundered, as it was always likely to do; but the attempt along the way to manipulate events in the wished-for direction caused a great deal of the council's dirty washing to be shown in public, which provided further evidence for a series of press articles about the dictatorial style of the council's leadership, and dealt a blow at least for a time to effective working relationships between the council and one of its key economic players that the council would normally be very keen to work with. It must also have cost the council a certain amount of money, in terms of wasted staff time. Whilst all of this was described internally as a planning failing, to my mind it was essentially a failure in the council's central policy processes; and in particular, it was a failure to question sufficiently strongly at the appropriate time the practicality of achievement of the wishes of the council's political leadership, because those wishes appeared to be treated as if they were unchallengeable.

The local press did not manufacture this story. Even without any press coverage at all, this would have been a very difficult case of a clash between planning policy advice and the known views of the council's political leadership, and then had the views of the leadership prevailed an equally difficult appeal would have had to be defended if no other solution could have been found. What the press coverage undoubtedly did do, however, was to make what would in any event have been very difficult, several times more so. Not only did it appear to me to have the effect of casting positions in concrete, thereby making impossible the process of exploration for solutions that all parties might have been prepared to accept, but also it became an issue about the internal workings of the council as well as about the merits of the arguments in their own terms. The fact that light is not thrown on such matters all that frequently undoubtedly contributed to the coverage that was given, as did the fact that some of this coverage made use of leaked documents in ways that enabled stories to be labelled 'exclusive'. I felt that all this went beyond what a local government chief officer should have to accept in the course of duty, even allowing for the fact that I was working for a very political authority in a city with a strong media presence. The distinction between what is acceptable and what is not acceptable is a difficult line to draw, but for me coverage that went beyond the issues and dealt with officers as personalities was pushing hard at this line. On the other hand, it is impossible to deny that all this must have looked to the local press like a very good series of stories; the accumulation of material about a range of issues that it engendered went well beyond what one would normally have expected when dealing with disagreements over a planning policy matter, and the momentum of the story certainly continued to build. This was a case, therefore, where press coverage became a major force in its own right; and unfortunately for the planning process in Manchester, it was a case where a relatively straightforward matter of planning principle (the policies in the development plan stand until it is demonstrably the case that other material considerations should prevail) was swamped by a lot of extraneous but probably more interesting material about how the political process worked and how member–officer relationships were handled.

Conclusions

In many ways, the story of planning in the inner city of Manchester as presented in this chapter is not a story of success. The necessary national policy frameworks have not been there with the consistency and the continuity needed. Locally, the focus has been on what has practically been possible, and this has been considerably less than the real need. Overall, the circumstances of many of the people living in the inner city have probably been worsening, and in recent times a public financial resource position which has focused largely on small numbers of targeted areas without tackling many of the wider issues has probably contributed to this problem of overall decline, despite some notable local achievements in places like Hulme. The planning service too has struggled at times to achieve everything it wanted to do, and has had to focus much more on what it has been able to do. It has had to cope with a legacy of 'planning failure' in the inner city, grounded in the 1950s/70s redevelopment phase, which from time to time became an issue capable of being exploited by political elements, apparently in order to attack the department. This is thus a story of struggle against major difficulties, rather than in any sense a case study of major successes.

And yet, it would be quite wrong to dismiss what has been achieved because it hasn't solved the inner city's problems, because it was never going to do that. This failure to live up to an unachievable standard does not negate the worth of large numbers of local projects carried out successfully with a high level of involvement of local people, of quite major achievements in the field of environmental improvements, and of the development of a series of community contacts throughout the inner city which have meant that the department has become increasingly better attuned to the needs of its customers in that area. If the 1950s/70s period of comprehensive redevelopment via a housing-led juggernaut which didn't involve local people was 'planning failure', then a more incremental approach which emphasises high levels of community involvement and the use of public expenditure to do worthwhile and widely supported things at least represents a process of improvement, even if the reputation lingers long after the event. In addition, the role of the planning service in contributing to the launching of major initiatives in a variety of ways has been an important one, as the Hulme City Challenge experience illustrates, and it is probably fair to say that circumstances in the inner city would have been worse had these sorts of opportunities not been grasped.

What is really needed, however, is the rediscovery of the inner city as a major issue, not in terms of the 'pepper-potting' of individual high-profile initiatives but through a sustained commitment to an attack on urban deprivation. This cannot just be done at local level; it needs a national commitment with a promise of sustained policy and project support which City Pride might have led to but as yet shows no real signs of doing. Whilst it is almost certainly the case that the various parts of our cities will for a long time to come exhibit degrees of inequality, the nature, the scale and the trends inherent in what is visible now are exceedingly worrying in terms of their long-term implications for our cities. The ever-more polarised city is surely not a goal to strive for, but it appears to be the outcome emerging from what is actually being done. We need to tackle this much more firmly both locally and nationally; the inner cities need to move back up the agenda. If we don't do this, there must be a high risk that we will reap the harvest in the form of permanent alienation, high levels of social disorder, a continuing very high support burden to be paid for out of the public purse, and large numbers of wasted or underachieving lives.

Part of this process must therefore involve harnessing the knowledge, experience and commitment of as large a number of people as possible to the goal of regenerating the inner city. This must include the residents of the inner city, and a planning process with a community basis to it has an important contribution to make. City Pride's great strength was that it involved a wide range of people from many (but mainly professional or business) walks of life in thinking not merely great thoughts about the future of the city but also about how those could be translated into reality; it achieved the process of getting the *cognoscenti* talking to each other more effectively and on a wider basis than had previously been the case. This is a necessary, but almost certainly not a sufficient, condition of effective inner-city regeneration. What needs to accompany this is a focus on the sorts of practical steps that can be taken in the various localities of the inner city to improve the quality of people's lives day in and day out, in which the people of those areas themselves have the opportunity to be active participants. This must be grounded in those localities and those people; it must largely be bottom up rather than top down. This Utopian view is some way away from what we have seen in this chapter; the shifts in policy regimes, in resourcing, in approaches to governance, and in priorities would have to be very great before it could be said that we have moved from where we have been to this changed focus. In reality, this will have to be fought for not only politically but also with the people of the inner city themselves, because they need more hope than I judge many of them actually have today.

This means a greater willingness to recognise that a range of approaches will need to be pursued together. Diversity will be an important component of this. There isn't one set of right answers, be this a belief in property development-led urban regeneration, a focus on prestige projects, a fervent belief in particular sets of approaches to the design and layout of physical development or whatever the latest big idea is from a politically dominant source. All these things have their place, alongside a range of other perspectives which are not always congruent with any ruling set of views. The process needs to open up and to embrace variety. We need to rediscover the Geddesian emphasis on folk, place and work in the inner city, and to do this in partnership with its people. The Manchester experience probably shows more about why this is needed than about how to do it.

None the less, the Manchester experience points at two major areas of work which are likely to be central to the translation of these broad thoughts into valuable practical reality. The first is that we need to find ways of harnessing the progress that can be made at the larger scale in reinventing the economies of our cities, as Chapter 6 has demonstrated, into direct benefits for inner-city residents. The positive trajectory of what has been described in Chapter 6 has nevertheless had insufficient effect on the everyday lives of the majority of people in the inner city to prevent their circumstances overall from deteriorating in several senses, and this is something that needs to change. More research, more policy analysis, more testing of what is possible, more experimentation; all these things need to take place to try to ensure that the residents of the inner city participate more fully in the 'good news' that work on the economic bases of cities can generate than has been the case to date. The second major area of work relates to the development of more localised perspectives, which unite contributions from inside a local authority (planning, housing, economic development, education and training, leisure in all its forms, for example) with inputs from other public and private agencies and local people to focus on the needs of an area and its people. To date, this has happened mainly when significant amounts of public resources have been available to pour into an area, as has been the case with Hulme; but at the same time many other areas whose people may have many difficulties deserving of such concentrated attention have not received this because large-scale new public resources have not been available. Hulme is being redeveloped but much of the rest of the inner city has continued to struggle; and this same pattern will be visible in many other cities. The major need now is to find ways of tackling these other areas, probably with far lower resource levels than in the high-profile major regeneration initiatives, but with no less a commitment to the spirit of local partnership, sustained over a period of time by at least a minimum guaranteed budget.

This will require local authorities to look hard at their administrative and financial frameworks, to satisfy themselves (and the people of these areas) that they are getting the most out of the resources already being put into these areas. When resources are scarce it is necessary to look critically at the scope for reusing the resources that already exist, rather than relying on what is ultimately a zero-sum game of shouting for more. None the less, there is enough accumulated experience of locally based working, and enough evidence from partnerships of all kinds at the local level, to suggest that an approach of this nature for the majority of areas coexisting alongside the kinds of major initiatives that secure national finance on a competitive basis (if this is to continue) for a minority of areas, has far more to offer than a continuation of apparently inevitable decline across too much of our inner cities. Something like this needs to be attempted if we are indeed to move away from the 'pepper-potting' approach to inner-city policy that has characterised the past few years.

The need to look at financial frameworks and at access to resources has been accentuated in recent years by the growing importance of European funding as an element in regeneration (Garside and Hebbert, 1989, pp. 173–90), particularly at a time when other sources of funding to local authorities were drying up (Stoker and Young, 1993, pp. 151–78). Plans and programmes designed to access this sort of funding have traditionally had a strong central government component to them (see for example the programme for the Manchester, Salford and Trafford Integrated Development Operation; United Kingdom Government, 1988), but for the two three-year programme periods 1994/99 within the context of a regional economic strategy (PIEDA, 1993) the Greater Manchester authorities tried to assemble a Greater Manchester strategy and programme based upon the needs of the area and agreed with local partners (Association of Greater Manchester Authorities, 1993). I spent a period of time as leader of the 'crash team' drawn originally from five of the Greater Manchester authorities set up to prepare this programme, which achieved the objective of obtaining local partner agreement but was not ultimately able to carry with it the government at national and regional levels because of the view that a single programme document was needed for the whole of the eligible area in the north west, which was wider than Greater Manchester. None the less, the effort put into this reflected the scale of available European resources and their relative importance compared with other sources of 'uncommitted' money. English Partnerships (1996) have calculated that, in all the eligible areas in England, European Regional Development Fund resources to the tune of £2,775 millions are available over the two programme periods between 1994 and 1999 inclusive. This works out over this period as a yearly average of over £460 millions, which is approximately twice the current annual budget of English Partnerships. This puts a scale on the importance of European resources, and it perhaps also serves to remind us that if the European Union expands into the former Iron Curtain countries from the year 2000 onwards this level of resource may not be available to us again. This experience of working on European programmes was also an important component in the development of my own thinking about the regional level of activity as a means of securing resources for inner-city regeneration, causing me to reflect whether the four contiguous conurbations of Merseyside, Greater Manchester, West Yorkshire and South Yorkshire with their concentrations of essentially similar urban problems weren't a better basis for thinking about an appropriate region than was the traditional concept of the northwest (Kitchen, 1996c).

Notes

1 It should be noted that the inner-city definitions used for the purposes of the material illustrated in Figures 7.1 and 7.2 are slightly different, because Moston was included with an inner city cluster in the 1993 study but was not defined as part of the inner city under the Inner Urban Areas Act 1978, whereas Whalley Range is shown on Figure 7.1 as an inner-city ward but was included with a non-inner-city cluster in the 1993 study.

8

Transportation planning

Introduction

Most people would probably accept that what is done in terms of transportation decision-making is likely to have a very considerable impact on the core city of a large conurbation. And yet, some of the trends of recent years have made the concept of transportation planning in a major British conurbation a harder one to turn into meaningful reality than arguably it has ever been before. Whatever the arguments in their own terms, bus deregulation, the fragmentation of British Rail in preparation for privatisation, and the reduction in the effective powers of passenger transport authorities and executives are difficult to see as policy initiatives designed to strengthen the transportation planning process in our conurbations. At the same time, it was still necessary for public authorities to try to grapple as best they could both with the direct problems associated with operating transport systems and with their indirect consequences. This chapter is about some of the tensions to which these efforts have given rise.

The chapter begins with an overview of some of the key information about the city's transportation system and the jobs that it does, commenting as it goes along on some of the local policy issues and difficulties which have arisen and occasionally on those that might have been expected to arise but haven't. Two deliberately different case studies have then been chosen from amongst the wide range of issues that could have been handled in this way, in order to illustrate aspects of the transportation planning process that raise different issues from those that are usually debated when local transport issues are discussed: the Channel Tunnel and the Section 40 process, and the expansion of Manchester Airport. Finally, some conclusions are drawn which seek to link these discussions about transportation policy issues with the wider questions about planning policy and about the nature of the planning job which are raised elsewhere in this book.

Box 8.1 Main transport trends and issues in Manchester, 1995

1) The trend in the modal splits for all journeys to the city centre has been moving in favour of the private car for some time now.
2) The volume of traffic entering the city centre grew between 1984 and 1989, but fell back slightly between 1989 and 1994. The private car showed a steady rise in its share of these totals throughout this period.
3) Private vehicle ownership appears to increase by broad bands as residential location moves away from the inner city. So, vehicle ownership levels are lowest in the inner city and highest across the rest of Greater Manchester outside the City of Manchester. This has quite major policy implications, because broadly it means that an emphasis on public transport favours inner-city residents who tend to be public transport-dependent and an emphasis on the private car favours suburban residents who are much more likely to be car owners.
4) Private vehicle occupation rates for cars entering the city centre at morning peak times appear to be falling.
5) Off-street car parking spaces (ignoring private residential parking) in the city centre have been growing in total, but their balance has been changing quite dramatically in favour of public short-stay parking and against public long-stay parking as a deliberate act of public policy designed to encourage visits for various purposes such as shopping, but to discourage all-day commuter parking.
6) Bus mileage has been growing, but passenger numbers have been falling. The effect of this dismal combination is that the average numbers of passengers carried per bus-mile travelled halved between 1985 and 1994, to my mind as a direct consequence of the disastrous government policy of bus deregulation initiated in 1986.
7) Metrolink has been a great success in patronage terms, but the rest of the railway system has undoubtedly struggled.
8) 1991 Census figures show that travel patterns to jobs in Manchester vary considerably according to whether or not people live outside the city. If they do, the car is the overwhelming majority mode of travel to work.
9) Manchester residents who work outside the city have a pattern of travel to work which is very similar to that of suburban commuters who travel into the city to work.
10) The solution to these problems is not simply to build more Metrolink extensions, although this would certainly contribute to the range of solutions needed. This is because, even if all the currently proposed Metrolink extensions were to get constructed, over 70 per cent of Mancunians would still not live within 10 minutes walking distance of a Metrolink station. Transport policy needs to be developed in the round with a major and continuing role envisaged for the bus, which is still responsible for the greatest number of public transport trips.

Transportation issues and problems in Manchester

In the interests of avoiding the presentation of a large number of statistics, Box 8.1 summarises the key transportation trends and issues affecting Manchester as of the middle of 1995. The basis for this summary is some figures contained in a paper I published as a result of a presentation at a Town and Country Planning Association Conference (Kitchen, 1993b), many of which were updated and added to for a special private meeting of the Labour group of the City Council to discuss transport policy which took place on 7 June 1995.

The major policy implications of the trends and issues summarised in Box 8.1 are as follows:

- Whilst transportation is about much more than access to and movement within the city centre, it should be clear why transportation issues frequently form such a critical element of policy discussions about the city centre. Its economic survival depends upon the effective operation of the transport system, not just in moving people to and from it for work or other visiting purposes but also to enable its various functions to be effectively serviced. At the same time, the sheer volumes of activity that this gives rise to in terms of people and goods movement are a very major contributor to its environmental problems, and the balance between meeting the centre's economic needs and improving its environment in order to maintain and enhance its attractiveness to its customers is one of the most critical (and one of the most controversial) of the current planning issues in Manchester. This has taken several forms, which illustrate some of the tensions which can exist between 'economic' and 'sustainability' perspectives, but at the heart of them has been a debate about how much restraint over private vehicle access in the city centre should be exercised. All this is examined in more detail in Chapter 9; see also Whitelegg (1993) for a presentation of these issues in a wider context.
- Equity issues ought in the light of the available statistics to be a more significant component of these debates than they yet have been. The available information shows unambiguously that the inner-city population of Manchester is predominantly non-car owning, and that the population of the surrounding Greater Manchester area is predominantly car-owning. Policy that starts from the desirability of unimpeded access by private car, whatever its virtues may be in terms of (for example) supporting the economy of the city centre, must therefore put the interests of the car-owning non-inner-city dweller ahead of the interests of the non-car-owing inner-city dweller. Indeed, in the fairly recent past it has been much worse than this. Improvements to radial roads as part of comprehensive redevelopment and slum clearance were designed to speed up journeys into the city centre by creating more highway capacity, but this was often at the expense of the inner-city dweller in at least two ways. The first was through the creation by the road schemes of larger physical barriers to lateral movements than had been there before. The second was as a result of the loss of

large numbers of the small businesses that had previously lined many of Manchester's radial roads and had often provided local employment. This was all facilitated by a method of justifying highways schemes which valued the time savings of business people (in those days, usually men) at several pounds per hour but the time losses of the housewife at a few pence per hour (see, for example, Mansfield, 1970). This record, which is to be found in many other cities as well, might have been expected to have generated a 'pro-inner city' lobby in terms of transportation policy, and perhaps in particular a strong 'pro-bus' lobby given both the low levels of car ownership to be found in the inner city and the virtual irrelevance of the railway system to most of the transportation needs of inner-city residents. Far from it, however. Much more typically, Labour members (from whom these sorts of attitudes might primarily have been expected), and particularly those who were themselves car-owners, would often take the view that the real need was to try to ensure that the undoubted benefits of car-owning became available to the whole of the population, and that this process should not be inhibited by policies of restriction. The council's political leadership would tend on the whole to support a similar policy stance, although it would be more likely to come at this from the angle of support for the city's economy than from the perspective of the poor person's aspiration to have the same freedoms as car-owners. The combination of these sets of thinking, however, explains why in the early 1990s very little fresh pedestrianisation or even pedestrian priority measures were attempted in Manchester city centre. At the same time, the economic circumstances of many inner-city residents worsened, and the bus system deteriorated. Thus transportation relativities between inner-city residents and others widened rather than narrowed. This is perhaps another dimension of the tension between economic development approaches which emphasise action to strengthen the economic base of the city and socio-economic deprivation approaches which emphasise action to improve the living circumstances of the poorest citizens (see Chapters 6 and 7).

- Flexibility and adaptability are obviously critical to the future prosperity of cities, and this must mean in physical terms as well as in other ways. The largely grid-iron layout of much of Manchester city centre has shown itself to be very flexible; for example, during the period between 1990 and 1992 when a significant number of streets were closed or reduced in width whilst Metrolink tracks were being laid, the city centre, despite what was said in some quarters, did not grind to a halt. Transportation systems, however, seem in practice to be very difficult to match to this requirement for flexibility and adaptability. Fixed-link rail systems are by definition inflexible, and thus can only survive when sufficient volumes of business are available alongside or adjacent to where they are located, or can be fed into them. Bus operations, although on the face of it potentially much more flexible, seem often to be much less so in practice, and many bus routes in Manchester today are recognisably the same as the routes formerly operated by the old-style tram system despite all the changes that have taken place since then. It is perhaps the flexibility of the private car more than anything else that has seen it grow into the most important force physically shaping our 20th century cities. One of the greatest challenges facing our planning systems, therefore, is how to generate movement opportunities for people that can at least begin to offer flexibilities approaching those offered by the private car. We are clearly not going to be able to do this everywhere, and in any event the private car isn't going to disappear tomorrow, because for most people the individual benefits they feel it gives them outweigh what they understand about the collective damage it may be causing. But how this balance is handled will be critical to our transportation planning efforts for the future.

Of course, unexpected major problems can sometimes bring in their wake opportunities for change on a scale greater than might otherwise have occurred, at any rate over the same time period. In this context it may well turn out that the response to the IRA bomb in the heart of the city's commercial core touched upon in Chapter 6 may provide opportunities for the introduction of some radical changes consequent upon the redevelopment necessitated by that outrage.

The Channel Tunnel and the Section 40 process[1]

Section 40 of the Channel Tunnel Act placed a duty on British Rail to prepare a plan stating the measures that needed to be taken both to stimulate international services through the Tunnel and to increase the proportions of passengers and goods travelling between places in the UK and places outside it carried by those services. The Act also placed a duty on British Rail to prepare such a plan by 31 December 1989 and subsequently to keep it under review, and it also clearly implied that these things would need to be done in a consultative manner. This came about through lobbying work led by the North of England Regional Consortium (the officers and offices for which were provided by Manchester City Council) which resulted in this section being introduced as an amendment in the House of Lords stage of the passage of the Bill and its ultimate acceptance by the government.

My role in all this was triggered by the fact that I was nominated to carry a series of portfolios into the consultative structure for discharging these responsibilities: representative of the North of England Regional Consortium (NOERC), of the Association of Greater Manchester Authorities, of the Greater Manchester Passenger Transport Authority and Executive, and of Manchester City Council. To the extent that these organisations had any very clear view at all of what they wanted to get out of this process, at any rate at the outset, it is probably not too unfair to say that this was initially motivated by fear. The Channel Tunnel, by virtue of its location and functioning, was seen potentially as something else that might drag economic activity into the southeast, and even conceivably into the economic heartlands of western

Europe. The worry about this, and the fact that it might happen without there being any very effective public debate about it, was really what promoted NOERC to take the action that it did, and this meant that what we started off with in terms of these processes was an essentially defensive position.

The other main participants in the process were people like me acting as representatives of clusters of authorities from various parts of the region (led in the shire areas by County Council staff and occasionally elected members), representatives of private sector interests through bodies such as chambers of commerce, representatives of business interests very directly involved in the fields being discussed in the process, trades unions, and a small number of people from transport pressure groups of various kinds. In practice, at the smaller meetings this tended to boil down very largely to a leading role being played by the local authority representatives, although a small number of other people stayed throughout the process as consistent players in the game.

British Rail's position appeared to be relatively straightforward. They hadn't asked for this extra duty to be imposed on them, but given that it had been they could see real advantages for them in a regional consultative process in respect of the potential freight market, because such a process was capable of improving very considerably their understanding of the regional market that the Tunnel might open. Very simply, this stemmed from the proposition that freight was potentially a very profitable operation over the sorts of distances that the Tunnel would facilitate, provided this did not involve too many breaks of bulk (because this is expensive) and could offer to customers a reasonable deal in terms of speed and reliability. From this perspective, the north west as the largest regional market in an absolute sense outside London and the south east was a potentially lucrative proposition. British Rail's perceptions were wholly different in relation to passenger services, however. They saw the real business opportunity here as being high speed trains between London and Paris or Brussels, competing well on a time basis (city centre to city centre) with the competition by air, and offering the business market a package in terms of price and comfort that would be very attractive. The further away one got from London, the less potentially attractive these propositions appeared to British Rail to be, and the regional tourism market was not seen as a major opportunity because they did not believe that train services by themselves would add value to what was at present on offer in terms of British regions as tourist destinations for residents of western European countries. These perceptions on the part of British Rail (and this is my interpretation of what their preconceived positions were; I certainly don't recall anything as stark as this actually being written down by them) determined how they approached the Section 40 task.

The mechanisms that emerged fairly quickly for tackling the job specified by the Act were an occasional large-scale consultative meeting of a wide range of regional interests, smaller working meetings broken down by subarea and by type of activity, and the submission in writing of evidence to British Rail to enable drafts of various parts of the plan to be produced by and consulted on by them. For the north west, with no real tradition of working together at the regional scale between sectors, this led over a period of time to a degree of understanding that we needed to get together and to try to present some co-ordinated views if we weren't going to be swamped, and also a growing understanding that we needed to do this in a way that emphasised operating opportunities to British Rail rather than the fears and worries arising from the region's initially defensive attitudes. Of course, this was overlain by a series of more local positions taken up within the region, some examples of which were as follows:

- Cheshire's concern to get the Holyhead line electrified, both to support Holyhead's role as a terminus for Irish traffic and to maintain services along that line and Chester's role as a railhead for it.
- Merseyside's desire to see itself as the primary port for Irish traffic, bolstered by arguments about 'landbridging' (the idea that it may make sense for customers to reduce the length of shipping journeys from Ireland or North America, and therefore the time taken, by breaking bulk at Liverpool and transferring to the rail system for a trip to the ultimate destination via the Tunnel rather than take the longer route direct by sea to a western European port).
- Arguments within both Greater Manchester and to a lesser extent Merseyside about the appropriate locations for Channel Tunnel freight terminals.
- Fears of being isolated on the part of Cumbria and north Lancashire in particular, which took the form amongst other things of arguments about passenger services through the Tunnel coming down the West Coast Main Line through Carlisle and Preston rather than starting in the conurbations at Manchester or Liverpool.
- Worries in several quarters about long-term underinvestment in the West Coast Main Line, and various views across the region about what the most appropriate response to this situation would be.

What became clear fairly quickly was that the process of the region achieving some sort of consensus could not be one of expecting people to ditch whatever views they held about these sorts of issues, in favour of some higher-order views around which everyone could unite. Apart from the local political difficulties that would have been involved in attempting something like this, which by themselves would almost certainly have been on such a scale as to make such an approach undeliverable at least in the relevant timescale, there really weren't any mechanisms then available to us which could have facilitated the emergence of such an overarching set of views capable of commanding such widespread support. Consequently, the approach had to be one of trying to do the best that we could for the region starting from some very general propositions about ensuring that it did not miss out, co-existing with some usually much more specific local concerns of the types listed above. The hope had to be that we might feel our way towards a more coherent set of regional views without either falling out over our local differences or persuading British Rail that there was so little coming out of the consultative process from their perspective that it could be allowed to degenerate into nothing more than tokenism.

In fact, the process of consultation and discussion about freight issues went reasonably well, and although regional interests did not feel that they had got as much out of this process as they had hoped when British Rail's Section 40 plan appeared in draft, there was a general sense that some worthwhile gains had been made. The problem area related to passenger services. This was well reflected in the meeting arrangement patterns which emerged, where typically British Rail were represented by several people from various arms of their freight business at freight meetings but by one individual only at passenger meetings essentially sent to stonewall for the particular (and very limited) British Rail starting view of through passenger services from the regions. This had two primary effects:

- It put us all in the position of having to be case-makers for the region, and this helped not merely to clarify what we were all prepared to support but also to create a greater sense of regional unity in pursuing these ends.
- It made us realise that the process of lobbying the government, Parliament and other influential external organisations was becoming increasingly important the nearer we got to the due date for British Rail's draft plan and the more it became clear at any rate on the passenger side that we would get relatively little out of that document.

Increasingly, it was acknowledged that the process would carry on beyond the date of the submission of the Section 40 plan to the Secretary of State, with a new focus on the question of the government's reaction to the plan. One of the consequences of this effective extension of the process was that it made possible the concept of a consultancy study at the regional scale, setting down what was in the region's best interests and targeted not merely at British Rail but also at the government. The North West Channel Tunnel Group emerged as a loose confederation of regional interests to oversee this process, and under the energetic chairmanship of Ken Medlock (a septuagenarian Merseysider, who also chaired the board of INWARD, the regional promotion and inward investment agency) a consultancy study was carried out by PIEDA to form the basis for a series of lobbying activities on the part of the region as a whole. Perhaps the prime effect of this was that it added to regional cohesion, because the various constituent bodies had been prepared to fund the North West Channel Tunnel Group including the cost of the consultancy study, to accept in principle the PIEDA study so that it stood as a coherent document on behalf of the region, and to acknowledge that it spoke on behalf of the region when carrying out its lobbying role. The group also attracted a degree of recognition from central government, in the sense that it had a few meetings direct with Roger Freeman (then Public Transport Minister in the Department of Transport), who always gave the impression of listening very carefully to what the group had to say and of seeking out the views of the group on the questions that interested him. That having been said, the government ultimately seemed to take the view that their prime job had been to take receipt of the Section 40 plan, to listen to all the views that had been expressed about it, and to leave largely to British Rail's commercial judgements the questions about how much of it got implemented and how much further work might need to be done.

I had increasingly to drop out of the group's work during 1990/91 because of other pressures on my time as City Planning Officer, although the department continued to send a representative in my stead; but by that time the government's position was becoming clearer, and thus the recognition that a *laissez faire* attitude such as they were expressing was likely to mean that we weren't going to make much real progress was having an adverse effect on meeting attendances. None the less, I certainly regard this as one of the significant strands in the formulation of regional views which contributed to the greater ability of the north west in the early 1990s than in previous decades to agree things through the newly formed North West Regional Association. The legacy of the work on the Channel Tunnel from a north west and (in my case) conurbation perspective was therefore wider than merely the Channel Tunnel. However, the north west did get some freight facilities and services as a consequence of the process, but it did not get the through-passenger services from day one of the Tunnel's opening that it had been lobbying for. Indeed, at the time of writing the region has only got poorly marketed and often largely empty feeder services to London Waterloo, which was precisely the fear expressed by regional interests at Section 40 passenger meetings with British Rail.

Planning for the expansion of Manchester Airport

There is quite a long history to this which it probably isn't necessary to recite in full to get to the essence of what this relationship has been in practice. At its heart has been the view amongst elected members on the City Council that the expansion of Manchester Airport is in the economic interests of the city, both directly (in terms of the jobs it provides and stimulates) and indirectly (as a component of growing importance in terms of what the city has to offer to the wider world). This has not been moderated to any great extent by the recent emergence of concerns about sustainability or about more long-standing environmental concerns such as noise; the belief has been that the airport should be operated to the best environmental standards within this growth philosophy, rather than that the growth philosophy itself should be questioned.

In its own terms, the growth of Manchester Airport is undoubtedly a major success story. The rounded figures in Table 8.1, drawn from the Airport's Development Strategy, tell this story in very bald terms.

The Airport Development Strategy also suggests (p. 17) that the multiplier effect of the airport's activities across the region as a whole is such that the total job dependence in the region in the early 1990s as compared with the number of jobs on site was in the ratio of 3 or 4:1. If this ratio were to

Table 8.1 Passenger and employment growth, Manchester Airport

Year	Terminal passenger (millions)	On-site employment
1970	2.0	
1975	2.5	
1980	4.0	
1985	6.0	
1990	10.0	10,000
1995 (forecast)	15.0	15,000
2000 (forecast)	22.0	
2005 (forecast)	30.0	30,000

Source: Manchester Airport Company, 1993, Chapters 3, 6 and 7.

continue, the airport by 2005 would be responsible in its own way in the north west for a level of economic activity roughly equivalent to the number of jobs physically clustered in the city centre today.

Over the years, growth on this scale has required a lot of work to be done on facility development, with which the planning service has been intimately involved. The model that increasingly has been used to guide this process is that of sitting the Airport's Development Strategy inside the framework provided by the statutory development plan in the form of supplementary planning guidance. Two particular complexities have affected this:

- The organisational complexity. Until very recently the Airport's Operational Area boundary has extended outside the administrative City of Manchester and into Cheshire. This has meant that the planning framework has been not merely the development plan in the Manchester area (basically the Ringway Local Plan, the Green Belt Local Plan and the Greater Manchester County Structure Plan, until these were overtaken by the adoption of the Manchester Unitary Development Plan) but also the development plan as it relates to the relevant parts of Macclesfield District (basically the Wilmslow Local Plan and the Cheshire County Structure Plan). Inevitably, as with most statutory planning exercises because of the sheer time that they take, at any one point in time this has tended to introduce a degree of uncertainty because one or more of these components may well be at various stages of preparation or under review, and it has certainly meant that the development plan framework covering the Operational Area has been a patchwork quilt rather than a coherent, easily understandable whole. This was simplified in respect of the existing boundaries when in 1993 the boundaries of the City of Manchester were extended so that they coincided with the Operational Area, but it will arise again if the second runway development is approved (see below) because this will extend out beyond the new boundaries into Cheshire.
- The ownership complexity. Manchester Airport had been developed up to 1974 as a municipal enterprise. From 1974 to 1986 ownership was shared

equally between Manchester City Council and the Greater Manchester County Council. From 1986, with the abolition of the GMC, the airport has in effect been a private company wholly owned by all 10 of the Greater Manchester district councils and managed by a board comprising representatives of those authorities and full-time executive directors (in effect, some of the company's most senior managers). Share distribution in 1986 was on the basis that Manchester kept its 50 per cent ownership, and the GMC's 50 per cent share was divided equally between all 10 Districts, giving Manchester in total a 55 per cent shareholding. So the council was both majority shareholder in a private company and local planning authority for most of that company's Operational Area. Legally, of course, these two functions need to be kept separate; the council's decisions as local planning authority must not be, and must not be seen to be, influenced by its ownership position *vis-à-vis* the airport company. Practically, the manifest tensions to which this could give rise, especially since the City Council's Airport Board members tended to be drawn largely from its political leadership, spoke volumes about the need for a very close working relationship between the planning service and the officials of the airport company, and this tended to make us regard the airport company as a very particular customer of the planning service.

The forecasts summarised above show why in 1993 the airport company submitted a planning application for a second runway. Indeed, the forecasts up to 2005, if they were to materialise, would take the airport close up to the capacity of the runway configuration submitted. It should be said that the evaluation of the effective capacity of a runway is an inexact science since it is affected by so many variables. A level of activity on this scale would take Manchester Airport well beyond the level of activity of Gatwick Airport and into the position of being by quite some way Britain's busiest airport after London Heathrow. In these terms, therefore, the stakes are high, and the decision as to whether the Secretary of State having called in the airport's planning application is prepared to approve it is critical to the future of the airport and thus to its economic impact on the city and the region.

Since a great deal of effort had been put into trying to work out operating configurations and methods which minimised the environmental consequences of this (but not to pretend that there weren't any, which would have been foolish), we were certainly prepared to accept as part of our initial appraisal of the application that the economic arguments in favour of the expansion were the dominating matters, and we had been working closely with airport company staff for some time to try to ensure that this would indeed be the outcome by the time that the application came to be submitted. This meant that we appeared at the public inquiry into the proposal as a supporter of the proposed expansion, but also as the local planning authority anxious to ensure that the package of conditions that would be attached to an approval were in the best interests not only of the airport company as developer but also of all our other customers who would be affected by the development. In turn, this was undoubtedly helped by the agreement that was reached between the airport company and Cheshire County Council about the various things that the Company would commit itself to do in relation to Cheshire's interests in pursuing this development, which turned Cheshire County Council also into a supporter at the inquiry, although Macclesfield District Council remained in opposition. This agreement ensured that there was a package of conditions that could be recommended to the inquiry inspector which stemmed from more than what otherwise might have been seen as a too-cosy relationship on this matter between the airport company and the City Council.

The scale of the forecasts shown in Table 8.1 also shows why, in the search for economic activity, considerable attention has been paid in recent years to the scope in the Wythenshawe area in particular (because of its proximity to the airport) for sites which can accommodate functions which are a spin-off from airport expansion. This has raised the problem that has been seen elsewhere, of local people sometimes feeling that they are losing their green spaces to development of this kind but not on the whole getting the jobs that locate on those sites. Just as more attention needs to be paid to the issue of how inner-city residents can benefit directly from the growth sectors of the city's economic base, so more attention needs to be paid to helping the people of Wythenshawe participate more fully in the wealth that follows airport expansion. At the same time, it has to be acknowledged that Wythenshawe is the closest part of Manchester to the airport, and so to the extent that proximity is important in locational terms this is inevitably the part of the city that is going to be under the greatest pressure for sites for airport spin-off developments. The scale of this can be very considerable as well. For example, a large site immediately to the north of the airport was taken through the UDP process and was being negotiated, as I left the City Council, as a development partnership between the council and a major business park developer, which was capable when full at conventional ratios for such facilities of providing up to 4,000 jobs. Cities are rarely in the position where they can deal with projects on this scale in terms of job attraction.

One particular subset of the process of dealing with Airport expansion is the question of surface transportation arrangements. Historically, access to Manchester Airport had been overwhelmingly by private car, with public transport via buses and coaches performing a small-scale function only. In the early-1990s, however, after quite a long struggle, a direct access railway was provided as a spur from the Styal line providing services from Manchester Piccadilly and places west, north and east (and, incidentally, producing in its station one of the best modern buildings in Manchester); and after another struggle a southern spur was also under construction at the time of writing. This enabled the airport company to set a target that by the year 2005 25 per cent of all trips to and from the airport would be by public transport, and the

Table 8.2 Public–private transport splits forecast, Manchester Airport, 1995/2005

Year	Passengers	Share by public transport	Number of trips by public transport	Number of trips by private transport
1995 (actual)	15 million	10%	1.5 million	13.5 million
2005 (forecast)	30 million	25%	7.5 million	22.5 million
Change 1995/2005	× 2	× 2.5	× 5	× 1.67

City Council as local planning authority to say in accepting this target that it regarded this as a staging post towards higher figures in subsequent years. It was clearly a necessary thing for the airport company to do in terms of how it presented itself at the inquiry; with the debate about sustainability intensifying, it would have been very vulnerable had it not sought to move in this direction, and in any event higher proportions of private car access would raise difficult operational and car parking issues for the airport. None the less, the scale of what degree of change this commitment could involve is very substantial, as Table 8.2 indicates.

In an absolute sense, this shows a growth of 9 million private car trips over a 10-year period, which by itself raises major issues about matters such as highway capacity, land-take for car parking, and the pollution generated by road traffic. Relatively, however, by far the greatest growth will have to be achieved by public transport services; a five-fold growth over a 10-year period would undoubtedly pose a major challenge. This was one of the arguments that led to the proposal that a Metrolink extension to Manchester Airport from the city centre should be constructed, although in practice the main function of this may well turn out to be in terms of worker access. In turn this provided the opportunity for a loop-line through Wythenshawe to be argued for as a natural extension of the airport line, with the Planning Department in the vanguard both of the arguments in principle and of the discussion about alignments. This was an example of local benefit being sought as a result of an airport expansion proposal. A case for a Metrolink alignment to serve Wythenshawe by itself, no matter how strongly that would have been supported by Wythenshawe members, simply could not have been made successfully within the rubric of contemporary transport planning argumentation. It remains to be seen not merely what the inspector and then the Secretary of State make of all of this but also (if the scheme is approved) what actually happens in these terms. What does seem clear writing from a 1995 perspective, however, is that this debate has changed the way thinking about surface transport access to the airport could be achieved, compared with a position only a few years ago when the dominance of the private car was very largely unchallenged.

The statistics quoted above are also becoming of increasing importance in terms of the environmental debate about airport expansion. This can be looked at in several different ways. Historically, the main issue with an expanding airport has tended to be the problem of aircraft noise, although the improvement in engine technology in more recent times has tended to mean that the relatively straightforward equation that used to apply between growth and noise no longer does so. Increasingly, the major environmental issue associated with airport expansion is becoming the problem of surface transport movements, which need to grow in addition to general levels of traffic growth which are themselves regarded today as a major problem. The sheer scale of this at Manchester, and the associated land-use problems it causes such as the apparently inexorable demand for more land for surface car parking, are easy to imagine from the figures quoted above. Noise and direct land-take are important environmental issues, but increasingly, as airports expand in response to the growing demand for civil aviation, surface access will be seen as the primary environmental problem. Of course, it is possible to look at all this in two quite different ways. One line of argument, being pursued by its protagonists with increasing militancy, is that airport expansion for all these reasons is an unsustainable phenomenon which simply should not be allowed. The other is to say that, whilst the basic demand for civil aviation should be catered for, there are some very important environmental issues wrapped up in airport expansion decisions and more weight should be given to them. It will be interesting to see in this context whether, eventually, airport expansion goes the same way as have arguments about highways expenditure, where lobbying trying to achieve greater weight for environmental considerations has increasingly given way to the assertion (with apparently a considerable degree of public support) that demand should not be catered for because of the self-fulfilling nature of this process.

Conclusions

There is a very important sense in which transportation planning should not really be seen in isolation, but should rather be viewed as part of what is needed in order to make other things happen. Although the transportation industry is an important industry in its own right, its purpose is essentially instrumental; it moves people and goods around in order to facilitate other activities. Thus, transportation issues tend to surface as part of these other issues, and that is reflected in the structure of this book. So, for example, transportation figures as an element of all the chapters in this section about the major arenas within which planning activities take place, and transportation debates have been presented as part and parcel of the matters discussed in those chapters. It is in particular at the heart of many of the debates about the future of the city centre, which have been heightened by the IRA bomb.

The two case studies that constitute the bulk of this chapter both look at protracted processes in relation to massive pieces of transportation investment. The attempt to influence the provision of support facilities to ensure that the northwest region did not suffer from (the initial perspective) but, it is hoped, benefited from (the emerging perspective) the investment in the Channel Tunnel could not really be said to have been very successful. What it did achieve, however, was a greater sense in some quarters in the region of the potential importance of the regional dimension in decision-making and of the value of regional interests in getting together and agreeing some propositions that can be put forward collectively, even if these have to sit alongside continuing local disagreements that need to be respected for what they are. The legacy of this was probably a contributing element to the further development of regional activity in the early 1990s, although that mainly happened because of the growing importance of the regional scale of thinking which dominates European Union resource allocation processes.

The second case, to do with the expansion of Manchester Airport, is about a continuing relationship with a very powerful customer that is custodian of one of the city's major resources. Here, what is changing is the nature of the debate itself, although probably not as much as some people might wish. Both these cases involve the deployment of a considerable amount of specialist knowledge and understanding, in combination with a changing degree of awareness throughout the lives of the cases of how to relate them to the planning system and its products. Here the planning role is a relatively small part of a multidisciplinary and multiorganisational effort, although it can sometimes be very important in terms of raising questions and promoting debate. Both these cases in a sense also sit outside the mainstream of transportation planning work as it affects the quality of people's lives on a daily basis throughout the city, which is more to do with the daily operation of a transportation system designed to help people to move around the city to work, to shop or to carry out their other functions than to do with the Channel Tunnel and the development of Manchester Airport. None the less, they do show something about the range and extent of necessary planning activities that need to be resourced and that need to be handled with as clear a view as is possible about the positive outcomes that may be achievable, despite very real difficulties and very real limitations on what the powers and functions of planners effectively are in these types of situations.

One of the dilemmas in all this is the uncertainty about how much the planning system is capable of achieving by itself in helping to manage some of the adverse consequences of transportation activities. This relationship is a central element of the following chapter about sustainability and Local Agenda 21. It also now plays a key role in government policy in the form of PPG13, which attaches considerable significance to the role of the planning system in limiting the need to travel through its decisions about the locations of land and building uses. The links between land use and transportation are at the heart of trying to understand how cities function in a physical sense, and yet the ability of the planning system to intervene constructively in these relationships is limited (to my mind very considerably) by at least three forces: the incremental and often long-term nature of many planning decisions, as it is hoped this book as a whole has illustrated; the moves in Britain away from significant direct public controls over large areas of transport operation, such as rail privatisation and bus deregulation; and the continuing love affair with the private car in a market economy, which means that a major job remains to be done in seeking to change people's perceptions and behaviour. This book, it is hoped, has thrown some light on how one British city in the late 1980s and early 1990s was struggling, largely unsuccessfully, to grapple with some of these issues. It raises major questions about the extent to which some of these matters are capable of being handled without revisiting the balance being struck in this policy field between the stick and the carrot. My experience is that people's behaviour will not change by lecturing at them. If we want people to use their cars less, we will have to provide satisfactory alternatives that are acceptable in terms of price, reliability, convenience and comfort. But we may also be able to influence the way people perceive these alternatives by making the use of the private car, particularly in critical locations such as city centres, much less attractive through a variety of measures involving pricing, policy and physical restrictions. It may take both a bigger carrot and a more powerful stick to achieve this objective, and that will not be easy for any government whether national or local because it will be unpopular (at any rate initially) and expensive. The real question, however, is whether we can allow the trends summarised at the beginning of this chapter to continue.

Note

[1] This case study originates from a paper I delivered at the Managing the Metropolis conference in Salford in September 1989 (subsequently published as Kitchen, 1993a).

9

Sustainability and Local Agenda 21

Introduction

The concept of sustainability as a component of planning thought is very old. It is possible, for example, to argue that Ebenezer Howard (Howard, 1946, first published in 1902) and Patrick Geddes (Geddes, 1968, first published in 1915) both had very clear views about this, expressed in the language of their times but with a message that is timeless, even if it could also be argued that some of their prescriptions might not have produced sustainable outcomes. It is probably much less easy to argue that sustainability has always been a significant component of British planning practice, and that we have simply been waiting for everyone else to catch up. Whatever the merits of these arguments within the planning community, however, world recognition that there are some major issues here which need to be tackled is relatively recent at any rate in terms of its translation from polemical or philosophical writing (for example, Vickers, 1983) to mainstream respectability. For example, only very recently have we seen the emergence of material purporting to draw together good practice in this field (Department of the Environment and Department of Transport, 1995).

The classic definition of sustainable development in recent times is that contained in the report of the Brundtland Commission of 1987: '. . . development that meets the needs of the present without compromising the ability of future generations to meet their own needs.' Probably more useful operationally, however, is the attempt that was made in the United Kingdom Local Government Declaration on Sustainable Development (United Kingdom Local Authority Associations, 1993, reproduced in Manchester City Council, 1994b) to turn this into a set of key objectives that should guide action:

- Staying within the capacities of the natural environment while improving the quality of life.
- Offering our children opportunities at least as good as those available to us.

- Ensuring that the poorest and most disadvantaged in society are not, by reason of their poverty, forced to ignore the obligations of sustainable development.
- Integrating environmental policy objectives with social and economic policies.

There are two very important dimensions that are clear from this set of objectives and which are already affecting action in places like Manchester, which are not self-evident from the Brundtland definition:

- This is about a process and not about an end-state. It is about establishing a direction we want to go in and testing individual situations against that direction, rather than about declaring a series of absolutes.
- It is about finding a balance between community, economic, environmental, political and social factors that has constantly to be reviewed, rather than about a set of dominant environmental or ecological nostrums.

As a conscious process, this is very new; indeed, it could not really be said to have obtained much general impetus until the Rio Earth Summit of 1992 and the adoption there by many of the world's governments of Agenda 21 (Quarrie, 1992). Even then there is some dispute about how much real difference all this has made in practice. To say that as a process this is new is not to say that individual and important decisions made in the past were not good decisions in terms of sustainability. For example, the 1945 Nicholas Plan for Manchester championed the cause of clean air. But it took decades of patient work through the promotion of smoke control orders and the elimination of worst practices to produce a city where complete coverage had been achieved. One of the immediate visual consequences of this, through the building clean-up programmes that followed smoke control work, is that people discovered that Manchester was not the black city wonderfully portrayed in the impressionistic paintings of Adolphe Vallette in the early years of this century, but was actually a city with buildings of many hues. Ironically, some of this is now under threat from the pollution generated by the growth of road traffic, which is discussed in more detail below. The other more recent example of sustainable activity not originally promoted under the banner of sustainability is the introduction of Metrolink as a much-improved public transport system. The sustainability arguments that can be advanced in its favour are not lost on the people now promoting Metrolink extensions, however, and these will feature much more prominently in the future.

This chapter is about how Manchester City Council has begun to dip its toe into these waters. It is also about the contextual difficulties that have arisen in seeking to do this, which have almost certainly hindered progress so far and will continue to do so in future. It begins with a section on how far Manchester had reached by early 1993, and links this to the city's role as host city for Global Forum '94; and to a discussion of how important this was not just as an event but as a catalyst. This is followed by an examination of the work that was going on in-house to promote thinking about sustainability and to work towards the achievement of a Local Agenda 21 statement for Manchester by the end of 1996, and a discussion of that process. Finally, there is a discussion of how these concerns and processes were beginning to impact on 'other' policy areas in the city, a form of language that is itself redolent of the problem, since according to the propositions advanced earlier about the nature of sustainability, it should not be a question of 'sustainability' work and 'other' work as two distinct categories but rather one of sustainability as a concept influencing all the council's thinking. At the time of writing, all of this is ongoing. This is a general problem with a book of this kind, but it is a particular problem with the subject matter of this chapter because much of it impinges on very large policy questions in the city which cannot yet be said to have been resolved. This needs to be borne in mind in reading this chapter. Some of this material has also been presented elsewhere (Kitchen, 1996d) in looking at urban sustainability as a policy area where top-down views from government are not readily translated into bottom-up actions at the local level.

The position in 1993

In addressing this question, I had thought of reproducing the famous blank page chapter of Len Shackleton's football autobiography, when he faced the question of what the average football club director knew about football. While that would perhaps be a little unfair, it would not be wildly inaccurate. By early 1993, Manchester was simply not a significant player in any real sense in discussions and action about sustainability. Individuals certainly had views, but compared with the range and extent of discussions about this matter then taking place nationally, very little was happening within the City Council. This would certainly also be a view that external commentators on the council's action, for example local environmental pressure groups, would readily subscribe to. Indeed, it would not be unfair to say that the position in 1993 was that a Conservative government was out-radicalling the Labour City Council on this issue, and that Manchester had been slow to respond to a growing national momentum. Stoker and Young (1993, pp. 64–96) summarise the general position in Britain at that time.

So, what changed? My answer to this would simply be that the impending reality of Global Forum '94, as a major event scheduled to take place in the city, made leading elected members aware that they were facing a potential public relations disaster if, at the same time as hosting a major world event about urban and rural sustainability, their own house was in chaos. Members were aware that something fairly similar had happened to the city of Rio de Janeiro in 1992 on the back of the Earth Summit, and did not wish to see this experience repeated in Manchester. I am absolutely sure that many of them also had loftier motives than this, but this is certainly my answer to the 'what changed' question.

Manchester's role as host city for what turned out to be Global Forum '94 came about as a result of a bidding process initiated when I was Acting Chief Executive. John Major, the British Prime Minister, had promised at Rio that Britain would host the next progress-checking event, and during the second half of 1992 potential host cities for this event were invited to bid for it to the government. Several did, including Manchester. My recollection at that time is that the Manchester bid was primarily motivated by the understanding that the intention appeared to be to hold the event in early September 1993, which by helpful coincidence was a few days before the International Olympic Committee would be announcing the venue for the 2000 Olympic Games, for which Manchester was British bid city. Thus, this was a wonderful opportunity to demonstrate that Manchester could stage a major event, and an even more wonderful opportunity to get some positive publicity for doing this at a critical time. This was seen as a significant component in Manchester's pitch for the event. I am in no position to say what was cause and what was effect here, and again I am sure that there were some rather higher motives than this around. None the less, that is my recollection of what the driving force was at that time behind the process of bidding for what became Global Forum '94.

It is one thing to bid for an event of this kind, however, and quite another to deliver it. It fairly quickly became clear that it was impractical to do something on the scale envisaged for Global Forum in the time available. It was agreed with the government as a result that there would be a government-driven event in Manchester in September 1993 (the 'Partnerships for Change' Conference) which among other things would feed into work then taking place within government on the UK strategy on sustainability (UK Government, 1990; 1994) and then that the main Global Forum event would take place in June 1994. One of the consequences of this, incidentally, turned out to be a funding difficulty that was a problem throughout the life of Global Forum; the government put its money into *Partnerships for Change* (Department of the Environment, 1994a) and did not fund Global Forum (and, indeed, was largely conspicuous by its absence from its proceedings). This set a timetable within which, if the public relations objective was to be achieved, visible action had to be taking place in Manchester to demonstrate that the city was taking sustainability seriously within its own boundaries as well as running international events about it.

The 'solution' to the problem was to designate the Planning Department as the responsible department and the Planning Committee as the responsible committee, in this latter case with a change of name to the Environmental Planning Committee. This change was accompanied by a change of terms of reference, which among other things now included a clear statement about the Committee's lead role on behalf of the council on these matters and a responsibility to advise other committees about what they were doing and ought to be doing in this regard. Arnold Spencer, the Chair of the Committee, regarded this as a great opportunity to be grabbed with both hands, both because of his strong personal commitments in this field and because he could see that it was a chance to do things that might not otherwise be possible.

While I shared all this, I was also worried about our ability to cope with a potentially large and difficult area that would have to be tackled by utilising the staff and other resources that we had already got. Actually, this wasn't literally true, because it had also been decided, as part of this set of changes of responsibility, to transfer to the Planning Department the Fuel Efficiency Unit that was then part of the City Treasurer's Department, which subsequently became the Energy Management Group; see Figure 1.1. As it transpired (although we didn't know this at the time), this gave us an opportunity to generate a sustainability budget that was to prove a critical component in our ability to get things done (see below). I am sure also that this was an element in the later decision to combine the majority of the Environmental Health Department with the Planning Department. None the less, this looked like a large extra lump of responsibility and a short and high-profile political timetable against a diminishing staff resource base which was already proving to be a struggle when set against existing workloads.

On the principle that there are no such things as problems, but merely opportunities with varying degrees of visibility, we launched ourselves into this situation. The method used was that I and a few other staff members (from what was subsequently to be renamed the Sustainability Group) sat down with the Committee Chair, and in a series of brain-storming sessions tried to identify actions which the council could set in train which could be part of a sustainability programme that would start to deliver visible results quickly. The practice was that when something had been identified that might have potential, the officers were sent away to work it up as a coherent proposition and bring it back as such in written form to a subsequent meeting. By this process, we got to about 40 action points. The Chair's view, when looking at this set in the round, was that it would be much easier to market politically and locally the concept of the 'Manchester 50' than it would of the Manchester 40, and therefore we were sent away to generate 10 more. We then took the view that, if this was going to stimulate commitment to action across the council as a whole, other departments and committees needed to be involved in this as well and thereby to take ownership of it. So we decided that we needed to report what we were doing in the form of a challenge to others, both to look at the feasibility of what had been proposed to date and then to add other ideas to it. To forewarn my colleagues of what was coming, I did a presentation of this approach at a Chief Officers Management Team meeting prior to the emergence of the committee report which formally issued this challenge in July 1993. These processes, therefore, had taken roughly two months, from the political decision to give us these responsibilities in late May 1993, to the report to the Environmental Planning Committee asking them to endorse the approach, add ideas of their own and issue the challenge to other committees in July 1993.

The Environmental Planning Committee very enthusiastically endorsed what we were doing here at its meeting in July 1993, did indeed suggest some further ideas in what by most committee standards was quite a lengthy discussion, and agreed that all committees should receive a report on this issue including proposed further actions they could contribute during the September cycle so that the Environmental Planning Committee could take an overview of what was emerging during its October meeting. In a couple of cases, for good practical reasons, it took a bit longer than this, and in one case (the City Engineer and Surveyor's report to the Highways Committee) the requested approach was not followed but instead a critique of our work to date was presented; but generally speaking the response of my colleague chief officers was both constructive and encouraging, and they did indeed respond to the challenge. By these processes, the 'Manchester 50' became the 'Manchester 100'; and the mere fact that by October 1993 there were actually slightly more than 100 initiatives did not detract from the presentational advantages of the concept of the 'Manchester 100', which has been retained ever since.

Without going into the detail of what precisely constituted the 'Manchester 100', its breadth can perhaps be illustrated by the titles of the 11 clusters of initiatives that made it up:

- air and water quality;
- citizens challenge (basically, grants and awards schemes);
- Council challenge (basically, internal work within the City Council to review policies and procedures);
- energy saving;
- landscaping;
- litter;
- publicity;
- recycling;
- economic development;
- transport; and
- technology.

There are probably no great surprises in this list; it is probably the sort of list that people would expect to see emanating from a venture such as this. Equally, however, it does give a fair indication of the range of initiatives covered, and as can be seen from some of the other material presented in this book it was not afraid to be controversial. I have told the story of this process in outline in Whittaker (1995, pp. 5–8).

The fact that all this had been generated very quickly, and that the City Council could justly claim to have made a start with implementing some of these initiatives, meant that I could appear on behalf of the City Council at one of the workshops in 'Partnerships for Change' in September 1993 and talk about a credible (and it is hoped a creditable) process that was now under way in Manchester, which was one of the political imperatives that had been set. It did also mean, however, that if we were to be able to repeat this at Global Forum '94 there had to be some progress with implementing components of the 'Manchester 100' by then, especially since it was decided to document all this formally as an input to Global Forum (Manchester City Council, 1994b). In the event, the first annual monitoring report in October 1994 recorded real progress with 40 per cent of the initiatives, some progress with 30 per cent and little or no progress with the remaining 30 per cent. This demonstrated two things:

- This was clearly a process that would proceed at different paces for individual initiatives, and we needed to recognise this.
- If the concept of the 'Manchester 100' was to have continuing validity, we needed to be able to replace projects that had either been completed or that had proved undeliverable with new projects, otherwise eventually it would become the 'Manchester 0'. This was logical really anyway, because no one was pretending that there were only 100 sustainability projects to be done in Manchester and then it would be a sustainable city; but it meant that we had to think of this as a continuous process rather than as a one-shot programme.

Whenever I have told this story, typically four comments or criticisms have come back:

- *You have elevated to the status of action points some items that aren't that but are really commitments to consider, to explore or to discuss.* This is true. It also makes what we did like most other lists of this kind that I have seen, because consideration, exploration and discussion are sometimes the triggers to action and there is no shame in recognising that this is where we need to start. There are also many very specific actions in the list. Indeed, there is undoubtedly a relationship between this distinction and the first-year monitoring 'scores' reported above.
- *You have only done the easy things yet.* This is true. In my experience most policy processes work this way. Sometimes the easy things need to be done simply to demonstrate that things are changing. Often, some successes need to be achieved early on to maintain the commitment and momentum of participants, and to indicate to the reluctant that their turn will come. Sometimes, doing the easy things first can make the harder things that little bit easier. All these have been important in demonstrating that Manchester had started to take sustainability more seriously, and we do need to demonstrate this if we are to work successfully with others on these matters.
- *There is no logical basis for these actions, and no baseline statement against which to monitor progress.* This is true. Indeed, a rational planning model (see, for example, the advice of the International Council for Local Environmental Initiatives, in Whittaker, 1995, pp. 84–98) would almost certainly suggest that before embarking upon doing things we should have undertaken a comprehensive study to establish the totality of what needed to be done, some measure of relative priorities, and as precise an understanding as we could establish of what the current situation was in each

circumstance so that changes from that baseline position could have been monitored. Such a model would probably also require us to develop, as part of this expression of the totality of what needs to be done, some sort of vision of what a 'sustainable city' might be like. I for one would have had a problem with this, because I find the notion of what Manchester would be like as a completely sustainable city very hard to grasp. I am clearly not alone in this as Peter Hall's review of the literature notes (in Brotchie *et al*, 1995, pp. 16–19). I have no difficulty, on the other hand, in seeing how and where actions can be taken that will move the city towards a more sustainable state, and in envisaging a process that would embark incrementally on actions of this kind without knowing how far this might be taken. Thus, had we accepted that these matters needed to be tackled via the application of this 'rational planning model' approach, starting from where we were in May 1993, we would probably still have been writing that document now. And what we would also have been denying was the manifest evidence of our senses as human beings and the very clear feedback from many of our customers. The ordinary senses of sight, hearing, smell, taste and touch can tell us a great deal about what is not sustainable in our cities, without awaiting major areas of scientific inquiry. The everyday experiences of people in the city tell us a great deal about which elements of its life they think are unacceptable or undesirable. These things show us the sorts of things that need to be done and the sorts of directions in which we need to be moving; we simply need to be prepared to listen. They probably don't tell us 'how much' of this should be done, but should the absence of precise knowledge about this prevent us from doing anything at all? We decided that we needed to see baseline work and the identification of precise targets for monitoring purposes as parallel activities alongside doing things, rather than their being the precursor to action. The target was to bring this all together in the Local Agenda 21 statement for Manchester due at the end of 1996, rather than to worry too much about the sequence of activities up to that point. There is in all this, of course, a strong dose of Manchester pragmatism. There is also a perfectly respectable set of arguments to suggest that policy shifts often start simply from doing some things, rather than necessarily thinking the whole process through and having a master plan for action from the outset (Rein, 1983).

- *This is all about the actions of the local authority, and sustainability is about much more than this.* This is true. One of the principles of Agenda 21 is that at the local level local authorities should take a lead in securing sustainability strategies for their communities; encouragement to local authorities to this end can be found in Chapter 28 of the main document (see Quarrie, 1992). But local authorities cannot credibly be seen as offering leadership in these terms unless they are also seen to be taking steps to put their own houses in order. The council was determined that it should start with itself. The other dimension to this, of course, is that 'think global, act local' must be taken as meaning literally what it says; in other words, it isn't just a question of thinking great thoughts but it is also a question of doing local things. Local authorities because of the range of their responsibilities must undoubtedly be major doers in these terms, and so getting on with some of this immediately was important not only symbolically, but also in itself.

Global Forum '94

This is not the place to go into the ups and downs of the preparations for Global Forum '94. It is sufficient to say merely two things:

- The event that took place was a much scaled-down version of the all-singing, all-dancing jamboree that had originally been intended.
- More than once, the whole thing came close to being cancelled as it went from one crisis to another, and it remained a major financial headache to the City Council throughout. Indeed, as is often the way with these things, the rescue activity was in practice in the hands of a relatively small number of people, including Councillor Spencer as Chair of the Environmental Planning Committee.

My main personal involvement with all this came late in 1993, when I was advised by the Chief Executive's Department that it had been decided, in view of my growing responsibilities for and involvement in sustainability work, that I would now take charge of the process of preparing the Local Authority Key Sector component of Global Forum, with a small seconded team (ultimately of four people) from the Chief Executive's Department to do the day-to-day work. I did not view this with unalloyed pleasure. Very little preparation had in practice been done for this part of the proceedings. Thus the transfer of responsibility and staff to me looked rather like 'with one bound the Chief Executive's Department were free'. In addition, the bad news stories about Global Forum generally outweighed the good by a factor of several to one, and this message had been picked up by British local government, causing a widespread (but not universal) desire to put distance between most authorities and Global Forum and all its works, rather than a desire to help.

On the other hand, of course, there was a very important sense in which Manchester City Council was on show, not merely to British local government but to local authorities across the world. In addition, Local Agenda 21 work is intended to have a strong local government contribution to it, and half-way between the Rio Earth Summit and the due date for Local Agenda 21 statements to be ready across the world, there was an obvious need for local authorities to meet, to take stock and to exchange experiences. This was one of the things that Global Forum was intended to be, of course, but if we had decided not to proceed with its Local Authority Key Sector component, British local government would have found it difficult to provide an equivalent, at any rate within the time required. The decision was therefore taken that the Local Authority Key Sector event would proceed despite the difficulties and, that if it was going to be done at all, it was going to be done well.

Thanks to the seconded team, to those authorities which did support us with money and with advice, and to the staff who volunteered for a series of duties during the event itself, this is exactly what happened. Indeed, the success of the event both organisationally and intellectually exceeded our wildest dreams. We would never have believed that it was going to turn out as well as it did. Generally speaking, this experience was repeated across the range of events in the scaled-down Global Forum and in its associated academic conference. As a whole, Global Forum attracted around 1,500 visitors from over 60 different countries. Approximately 400 of these, from most of the countries of this range, attended the two main days of the Local Authority Key Sector conference itself, and some came only for this component; smaller numbers attended the specialist conferences the day before and the Commonwealth conference on the day after. This in itself was very gratifying (and, incidentally, made a worthwhile contribution to the city's developing tourism economy). The feedback that came from visitors meant that this was not the public relations disaster for the city that it could have been. Intellectually, in my judgement its products have made a valuable contribution to the burgeoning literature about sustainability (Dalton and Longhurst, 1994; Manchester City Council, 1995b; Whittaker, 1995), principally because they contribute a considerable amount of material from experience to the general material that is available (see Bartelmus, 1994; Blowers 1993; Haughton and Hunter, 1994; Williams and Haughton, 1994; Blackman, 1995, pp. 223–48; Gray, 1995; Gilbert *et al.*, 1996). Perhaps more than anything else, for those that participated there was the stimulus of exchanging views about broadly common endeavours with people from all parts of the world, and the opportunity provided by these experiences (and by the time away from one's desk) to reflect on what we were all doing. I tried to provide some quick reactions for the planning profession as a whole within a month (Kitchen, 1994), and my main conclusions both then and now were:

- In terms of the timetable, the world was supposed to be half-way towards the achievement of Local Agenda 21 statements, but in reality it was nowhere near this level of attainment although some individual areas had made very substantial progress. So, if this timetable is to be maintained, there will have to be a considerable acceleration over the second half of this period.
- Very well worked-out rational planning models are available to local authorities and have a role to play for those that are able to think and to act in this way. But for many authorities in many parts of the world who do not see themselves as being able to operate in this way in order to achieve something worth while within the Rio timetable, the pragmatic Manchester-type approach, which emphasises the possibilities of getting on with things straight away, has a lot to offer. My experience was that there was a lot of interest in what we were doing in Manchester from Global Forum delegates both from individual countries and from international organisations, on the basis that existing knowledge helps to define some clear directions and some specific

things which can be achieved. It was important in this context that people believed that it was possible to do something, and were not cowed by their perceptions of the differences between the rational model and where they were themselves into believing that nothing could be achieved.

- Sustainability did not end as an issue on 31 December 1996 when everyone's Local Agenda 21 statements were completed (if indeed they were). Statements are simply a starting point or, in some cases, the latest milepost along a road. Very few, if any, places could claim that they would be sustainable by the end of 1996, and indeed very few people (me included) felt able to define in practice what this would actually mean for their places. The emphasis thus needs to be on the process, and on making progress through a series of incremental steps, without worrying too much about reaching the final state of grace.
- A process that is about negotiated commitments between equal partners rather than about the decision-making power of local authorities is already providing new challenges to local authorities in terms of their behaviour. Generally speaking, local authorities are not as good at this sort of sharing of responsibility as they are at consultative processes designed to inform their own decision-making processes, perhaps because they aren't as practised at it.
- Without being territorial about this, Local Agenda 21 work is both a challenge and an opportunity for planners and planning. It is a challenge because stimulating and enabling roles need to be played, as well as traditional policies and practices rethought. It is an opportunity because it could help planning practice to rediscover and give added emphasis to its central intellectual concern with environmental matters.
- Perhaps the greatest challenge is the need to find effective ways of involving children in the Local Agenda 21 process as direct participants with their own rights and obligations. Children's views are often ignored, or presented (and perhaps distorted) through proxies such as teachers and parents. But more than anyone else for reasons of longevity, children have a stake in the outcome of Local Agenda 21 work. So if the concept of 'stakeholders' in this activity means anything it must surely apply to children as well. In addition, of course, children often have very acute awareness and understanding of environmental issues, and through their schools represent a major potential resource for undertaking environmental work. All these points suggest that finding effective roles in the process for children is a very important issue.

Local Agenda 21 work in Manchester

Getting on with the 'Manchester 100' was important in its own right and a useful statement to others about the council's willingness to contemplate changing its own policies and practices to help move towards a more sustainable city. By itself, however, it could only be part of the work needed to move towards having a Local Agenda 21 statement for Manchester and of securing the cross-sectoral commitments to action that would need to be at the heart of such a statement if it was to be more than merely some fine words. The big task, mainly for the period after Global Forum '94, was to set in train a process that would generate what was necessary, not by and for the council, but for the city as a whole. The line taken at early open meetings of interested parties was that this was not about the council as such but was about collective agreement and action. The council was prepared to play an enabling and (if wanted) a leading role in all this, but believed fundamentally that the process in the spirit of the Rio agreements ought to be an egalitarian one, and not the more normal council-dominated one that everyone was used to. Seeing all the potential partners in the process as 'stakeholders' was an essential part of the attempt to handle Local Agenda 21 work in this way.

Arnold Spencer took a leading role in all this. He believed very strongly that these words had to be given real meaning in Manchester, and he saw his leadership role very much in these terms. It would probably not be unfair to say that, as his position with the council's leadership became ever more embattled over a series of issues to do with the relative weight given to environmental considerations as against what he saw as the leadership's obsession with economic development and the increasing centralisation of decision-making within the council, so he saw increasing value in developing an external constituency in this way. As well as his own personal beliefs as a long-standing environmentalist, his experiences in dealing with Global Forum and all the debates surrounding what it should be about and how it should cover this ground, reinforced these commitments. His concern was to use the process of working towards having a Local Agenda 21 statement for Manchester as a means of changing not only some of *what* was done in Manchester but also *how* it was done.

Early meetings of what became known as the LA 21 Forum were dominated by at least three things:

- The difficulty that other people had in believing that the City Council might actually behave in ways that were different from those that they were used to.
- The difficulty that most environmental groups had in articulating what they actually wanted to see done, as distinct from using the forum as another opportunity to repeat views about what they were against.
- The unbalanced nature of attendance at meetings, and in particular the difficulty in generating much private sector interest in Local Agenda 21.

The way forward really had to involve working, at least at the beginning, with the people who wanted to participate on the issues that they wanted to talk about, with the council trying to act as gatekeeper of the process. At the same time, the council was exploring ways of broadening out to encompass those elements that it felt really ought to be part of the Local Agenda 21 process in Manchester but weren't as yet. For example, this involved convening separate meetings with private sector interests to explore ways of generating involvement. This process was still being struggled with when I left the City Council. The structure that emerged settled down as being a forum, a management group and a series of working/action groups. The forum, which meets at two-monthly intervals, consists of 50 people, with four from each of the following sectors and extra places for the local authority sector and the community sector:

- local authority
- academic/further education;
- community – residents/civic societies;
- community – campaigning groups;
- community – service-providing groups;
- health;
- trades unions;
- conservation;
- business/private sector;
- women;
- schools/youth; and
- other public bodies.

The management group, which also meets at two-monthly intervals but in the months in-between Forum meetings, consists of 15 people, one from each of these 12 sectors and then three officers, two of which are elected at large by the forum (chair and treasurer) and one (secretary) who is a paid official. In turn, the management group determined the membership of seven working/action groups, after the forum had agreed this basic structure for undertaking work following a considerable amount of debate:

- transport;
- health;
- energy;
- waste and pollution;
- greening and open space;
- economy and work; and
- education and consultation.

These seven working/action groups were intended to be the main motors of the process. They were to try to determine in each of their areas what basic actions needed to be undertaken and what information needed to be assembled in order to produce indicators of progress in each area. The intention was that a consultation draft of a Manchester Local Agenda 21 statement would emerge through this structure by early 1996, and each working/action group was to produce a first paper covering this ground for discussion initially by the management group in early autumn 1995. Whether this process will result in a Local Agenda 21 Statement for Manchester which really does move the city forward in terms of sustainability remains to be seen. But what can already be seen in this structure is how far the intention to ensure that the process is a genuinely collective effort has been taken. Whatever else may be said about this, it is not the City Council masquerading under another name.

Two other dimensions of this may be worthy of discussion. The first is the problem of scale. An administrative area such as that covered by Manchester City Council is probably not a 'natural' area for thinking about sustainability, and its currency should probably be regarded as being essentially political and administrative. There is thus a need to think about work at both the larger and the smaller scale. At the larger scale, the North West Regional Association (undated) had already produced a useful inventory of the need for environmental action in the region, which provided a broad context for work at the conurbation scale. This task was tackled by a piece of work undertaken by the Town and Country Planning Association and (mainly) staff at Manchester Metropolitan University. It was designed to produce common understandings about sustainability issues at the conurbation scale, which could in turn be picked up and used with a degree of consistency in Local Agenda 21 work in each district. At the time of writing, this piece of work is in draft form (Ravetz, 1995), but the need for it can perhaps be gauged from the fact that by late 1995 all the ten Greater Manchester districts had committed themselves to Local Agenda 21 and had made a start of at least some kind. At a smaller scale than that of the city as a whole, there is clearly a need to look at sustainability issues at the level of the locality. This probably interests most of the participants in the process more than its other elements. But, the way forward on this is not very clear, although it is likely to involve a recognition that uniformity across the city as a whole on many matters is not of overriding importance. Therefore funding to encourage local diversity through projects which are justified by their local significance and support needs to be found through the council's enabling role.

This leads on to the second dimension that ought to be discussed. The council's enabling role is in practice related to its ability to provide funding for new initiatives. This is what many groups participating in the Local Agenda 21 process in Manchester expect of the City Council. The doctrines of sustainability would suggest that it should not mainly be about new initiatives but about rethinking how we use the resources that we have already got.

The problem is that to get from where we are now to where we want to be does involve new initiatives, even if nil growth is the final result. Since there was no potential funder for this other than the City Council, we needed to find a source if in practice we were going to encourage new initiatives.

This was where the transfer of the former Fuel Efficiency Unit from the City Treasurer's Department came into its own. The political agreement reached was that extra savings generated for the council by the work of what was to become the Energy Management Group in the Planning Department would be split 50:50 between the council's general coffers and our new sustainability budget, provided that none of the latter was spent on planners. Over the three-year period 1994/95 to 1996/97 this produced a total of £1.2

million for this budget, growing progressively throughout this period, which was used to pay for (among other things) the following:

- Improved staffing for the Energy Management Group, because without this it wouldn't be able to generate the savings from which everything else followed. These people didn't count as planners for the purposes of the political agreement, because that isn't their disciplinary background.
- Staff resources to fund administrative support for the work of the Local Agenda 21 Forum and its offshoots, and also to help the education service promote Local Agenda 21 work in Manchester's schools.
- A pot of money to fund community-based sustainability initiatives at the local level.
- Capital resources to enable the City Council to carry out demonstration-type projects in the sustainability field.
- The City Council's contribution to the need for resources for new recycling initiatives, national resources having already been secured for part of the cost of this from the Department of the Environment.
- A contribution towards the publication costs of documentation recording the outcomes of Global Forum events.
- A contract with Leeds Metropolitan University's Centre for Urban Development and Environmental Management (CUDEM) to provide an independent monitoring function throughout the life of the Local Agenda 21 process in Manchester. This contract also enabled CUDEM to provide specialist inputs in the form of improved understandings of how LA 21 work was being handled elsewhere, and expert work in fields such as the development of the involvement of children and the development of our thinking about the relationships between local economic development and sustainability. It was important in these terms that a non-Manchester but relatively local capability was found for this work, because what was needed was an accessible non-stakeholder.

There is an argument which says that we should have been doing at least some of these things anyway without having to find new resources; and maybe we should. But my experience of the process was that the knowledge that real money was available to enable things like this to happen without having to fight the usual corporate battles had a galvanising effect on people's willingness to think in terms of new initiatives. As Bob Dylan put it, 'money doesn't talk, it swears'.

Policy conflicts

The message that never got across effectively to the council's political leadership during this time was that sustainability, among other things, needed to be about achieving congruency with what the council was doing in the field of economic development. The two needed to be seen in mutually reinforcing rather than mutually antipathetic ways. Because of the failure to establish this message, the view appeared to be that we could get on with sustainability work provided that it didn't clash with what was being done to build up the city's economic base. But as soon as any potential clash existed not only had the economic work to take precedence but also there had to be no suggestion that it might need to be moderated in any way by sustainability considerations. The two were placed in hermetically sealed compartments, one of which was deemed by the political leadership to be much bigger and much more important than the other. The issue on which these difficulties emerged most sharply was the problem of air pollution caused by road traffic, particularly in the city centre.

This issue really began to surface in a major way in Manchester during the second half of 1994. Local monitoring work in Manchester city centre, as in many other parts of the country, was showing that there was a significant problem of air pollution concentration, at any rate on certain days when the local weather conditions encouraged retention rather than dispersion. Two examples will serve to make the point:

- Nitrogen dioxide measurements for the period January–November 1994 showed that a city centre roadside location was typically recording twice the levels on average of three areas away from the city centre, and that the value of the 98th percentile not only continued this ratio but in all probability at 122 parts per billion failed to meet the EU directive limit value of 105 parts per billion.
- Carbon monoxide measurements showed that during the daytime on certain days the recommended air quality standard of 10 parts per million was exceeded alongside busy city-centre roads. For example, on 28 January 1995 (which was a poor day in terms of air quality generally because of weather conditions) this standard was attained or exceeded on average for every hour between 9 a.m. and 7 p.m.

This sort of information had led the government to go beyond some of its earlier views, and to suggest that the planning system could be used to deal with some of these issues by developing better relationships between land uses and therefore reducing the need to travel (ECOTEC and Transportation Planning Associates 1993; Departments of the Environment and of Transport, 1994). The government proposed in a discussion document in January 1995 (Department of the Environment, 1995) to introduce new statutory duties for local authorities to intervene more directly in such areas of concentration.

The political leadership saw this as worrying enough in terms of its potential effects on Manchester, given that in practice it would mainly be about taking action in relation to the city centre which was the jewel in the city's economic crown. What made it much worse locally was that the *Manchester Evening News* latched on to the publication of comparative information about cities by the government. These did indeed appear to show that on some occasions the air quality figures for Manchester were worse than elsewhere. The *Evening News* started running articles about Manchester as 'the most polluted City in Britain', and began campaigning for more action on this issue by the City Council. Arnold Spencer joined in, taking essentially the same tack, and arguing that essentially the problem was a failure to give due weight to this as an issue despite several existing City Council policy statements to this effect. He saw this as the logical continuation of the positions he had been adopting as part of the Local Agenda 21 process, and therefore necessary to retain faith with the participants in this process. This rapidly got publicised as a major row between Arnold Spencer and Graham Stringer, with the Leader of the Council taking the line that nothing should be done which might damage the economic performance of the city centre or indeed the wider reputation of the city, and that in any event the stories were exaggerating the problem. Bitter words were exchanged between them through the press, and I was told that this was nothing compared with what was happening between them in Labour group meetings and what each was saying to his own supporters. The local press lapped up these conflicts

Into all this stepped a new party at the 1995 municipal elections, at which Arnold Spencer was a Labour party candidate for re-election in his own ward. The new party was called Fresh Air Now (FAN), and in association with the Green Party they put up candidates in 30 of the 33 wards. Their campaign culminated in a series of stunts, such as for example sending out fake press releases on mocked-up press-office paper announcing the closure of a major city-centre street for environmental reasons two days before election day. The effect of all this on the electorate was negligible. In total, Green/FAN candidates took 3.2 per cent of the votes cast in the wards where they fielded candidates, and 3.0 per cent of the city-wide vote. Indeed, even within these very small figures there was a marked difference between the performance of Green and of FAN candidates, with the Greens on average taking three times as many votes per candidate as the FANs. All this may have suggested that there was no real public interest in the air pollution issue. Clearly it had not been capable on this occasion of shifting traditional party loyalties at local elections. Despite being returned in his own ward by a large majority, Arnold Spencer lost his chairmanship of the Environmental Planning Committee in the party and group elections that followed the municipal elections in May 1995. He was in effect cast into the political wilderness by being stripped of all his committee seats and all outside appointments. He still retained a role in the Local Agenda 21 process as Chair of the Forum, however, since this was in the gift of the Forum and not of the council, and he continued as a Forum member, not as a council representative, but as a representative of the further and higher education sector. The political leadership then took the issue up in the form of a 'task force' to tackle pollution. Its initial actions were targeted mainly at old buses, that as a result of deregulation were running a large number of dead miles in the city centre and were undoubtedly causing pollution, although they were almost certainly not the main cause of pollution in the city centre. At the time of writing, it remains to be seen what will happen to this initiative (which looks like the political leadership wanting to be seen to be 'in control' of the issue) and how it will link into the emerging Local Agenda 21 statement.

Behind this story of policy conflicts is clearly a very large issue. If it is indeed the case, as the Royal Commission on Environmental Pollution shows (1995, pp. 21–45), that traffic is the major cause of air pollution, then the solution will need to be an attempt to move towards more sustainable forms of transport which don't have such severe consequences. If Manchester's experience is typical, this will raise major concerns about whether such actions might not fundamentally undermine the economies of the cities whose problems are being addressed. Arguably, it is the concentration of activities in cities (and particularly in their centres) that give rise to these concentrations of problems in the first place. An agenda of this kind is bound to be very wide ranging, and certainly is not going to be limited to matters which can all be controlled locally. For example, at the very least such an agenda is likely to involve all the following:

- Applying both the stick and the carrot to attempts to encourage people when travelling to the city centre not to use their cars. The stick might include real reductions in car parking spaces and real reductions in highways capacity for private car use. The carrot might include better public transport services and facilities (such as improving radically the quality of buses) and fares policies that encourage their use.
- Using the planning system to address the long-term relationship between land use and transportation issues. This might be through policies which seek to intensify development in the inner city by bringing back into use land which is underused or unused; and by policies which turn down development proposals elsewhere. This would include those seeking to leapfrog the inner city precisely because such a policy framework existed. There will also be local authorities who might wish to suggest that they will offer a more car-friendly environment to economic activities in order to attract local moves.
- Producing better planning and financial arrangements for policies in relation to the conurbations, which switch resources behind policies of the kind described and take them away from policies that are not consistent with this thinking, and which also change the balance of resource provision in some programme areas (such as national transport expenditure by the Department of Transport) in favour of the conurbations.
- Mounting major public campaigns designed to change people's behaviour patterns, coupled with an aggressive use of the taxation system designed to support such changes.

Manchester's difficulties with a small fraction of these arguments, as documented in this case study, suggest that there is a long way to go before such an agenda could be regarded as having been widely accepted. This is at least in part because the frame of reference adopted by the political leadership, with its pre-eminent emphasis on jobs and economic activity, virtually precludes the possibility of accepting a sustainability agenda since it is seen (rightly or wrongly) as putting jobs and economic activity in jeopardy.

The importance of frames of reference in understanding policy discourses is discussed by Rein and Schon (in Fischer and Forester, 1993, pp. 145–66). Thus, sustainability debates ran largely untrammelled until Global Forum '94 was over because it was perceived to be necessary to ensure that the city's image emerged unscathed. But as soon as this phase was completed, the hegemony of the political leadership's views was reasserted; in effect, attempts were made to put the stopper back in the bottle. The political power struggle inside the council was won decisively by the political leadership over a relatively short period (July 1994 to May 1995). It could be argued that, outside the council, the constant evidence via the local press that this struggle was taking place had the effect of goading others to take more action than they might otherwise have, as the emergence of FAN at the 1995 local election perhaps illustrates. Inside the council, once this process had been completed, it rendered views alternative to those of the political leadership undiscussable, and in turn the fact of their undiscussability had also become undiscussable.

As Argyris (1993) shows, this is often a characteristic of an organisation that is not learning effectively, because it is denying itself access to much of the information and understanding that would help it to learn. There was no further debate since the dominant frame of reference inside the council precluded the need for such an activity. It had already decided what the answers were, and the leadership appeared neither to be open to the possibility of persuasion about alternatives, nor willing to attempt to win 'hearts and minds' outside as to the validity of its own views, which would of course have opened it up to forms of debate. This suggests that changes may have to await external events which challenge this frame of reference (such as political changes or new legislation), and may well be slow and gradual.

As Chapter 8 has already pointed out however, it is possible that one of the consequences of the IRA bomb in central Manchester will be to provide a catalyst for more radical transport policy changes than would otherwise have occurred.

It is also fair to say, however, that the intellectual argument for the sort of agenda outlined above has still to be won convincingly; thus it would be unfair to present the reactions of the council's political leadership merely as a reflex behind which there are no worries of substance. For example, the kind of agenda set out would be likely to be seen today as a relatively uncontentious expression of current conventional wisdoms about sustainability issues in relation to the problems of the core city of a large conurbation, and more attention would probably focus on the difficulties of implementation, rather than on the rightness of the prescription itself. Yet Breheny (in Brotchie *et al.*, 1995, pp. 402–29) adds to these worries about deliverability some real doubts about whether policies of urban containment are really as advantageous in terms of energy consumption as claimed, or if they are capable of affecting current urbanisation trends other than at the margin. He speculates whether conventional wisdom in this field isn't in danger of becoming a 'Canute-like' policy. Progress will clearly depend upon intellectual challenges like this being met, as well as upon political will.

Conclusions

This is a story of a process that has come a long way in a short time, and an area where some of the major internal political difficulties of the ruling Labour group have surfaced. In the early 1990s Manchester was not one of the authorities that came to mind when thinking about those councils taking a lead on sustainability issues. Yet by the mid-1990s it could make a respectable case for being considered in these terms. Much of the impetus for this was undoubtedly given by the city's impending status as host city for Global Forum '94, but much of it also was because one local politician used his position as Chair of the Environmental Planning Committee to take constructive advantage of this opportunity. The staff work in the Planning Department, which was essentially achieved by refocusing priorities, made an important contribution to this because it generated the material that challenged other people to help take the opportunities presented. All this was essentially pragmatic. There was no great period of introspection, no logical march through survey, analysis and plan, and certainly no very clear understanding of where all this was likely to take us. What there was, above all else, was a recognition of an opportunity and a determination to take it.

And yet, the political difficulties that the process has run into raise questions about what the effectiveness of all this will turn out to be in the longer term. Is this genuinely beginning to change the way a large local authority thinks and behaves? Or is it simply something that people were allowed to run with for image reasons, until it came up against the things that really matter to the political leadership, when the process (and the people) quickly got trodden on? Probably it's a bit of both.

For example, both City Pride and annual housing strategy submissions to government, as part of the Housing Investment Programme bidding process, have had quite a strong 'sustainability' feel to them recently. So at the very least the need for this kind of window-dressing has been recognised. Similarly, the ongoing nature of Local Agenda 21 work in the city forced the issue of traffic pollution further up the political agenda than it would otherwise have got, simply because of the contradictions that it exposed in what the council was doing and not doing. On the other hand, the members of the Environmental Planning Committee who came to a members' away day on sustainability in January 1995 were very clear that the main issue was 'mainstreaming', by which they meant the process of making sure that sustainability considerations became part of what the council was doing in all its major policy fields of activity. Thus, elected members were saying that they did not believe that the sustainability debate was having much of an impression on those people who were not immediately part of it. Subsequent events demonstrated the accuracy of this assessment. Post May 1995, the reassertion of the dominant view of the leadership about economic activity and jobs appeared to freeze the scope for making much internal progress with 'mainstreaming'. What it didn't do, of course, was to shut off external pressures on the City Council, and at least for a while these may be the best hope of those internal forces who want to see radical policy changes.

The big issue yet to be tackled is undoubtedly the need for sustainability not to be seen as antipathetic to work on strengthening the city's economic base, but for it to be accepted as encompassing it. This points up yet again the distinction between work on the city's economic base and local economic development work designed to improve the living circumstances of the city's most deprived residents. The latter would undoubtedly be seen as fundamental to work on sustainability, as the definitions quoted earlier in this chapter clearly show, whereas the former is still seen as being impeded by all those Green activists who have climbed on the sustainability bandwagon. The difficulties experienced with the question of vehicular movements of all kinds in the city centre, and the problems of air pollution they cause, illustrate this very well. This remains a major problem of perceptions which doesn't look as if it is about to be resolved, although it may be that a process of attrition will bring thinking on both sides of this argument somewhat closer together. It may also be that progress will be made through more attention being paid to the job-generating potential of green or environmentally friendly industries, which may turn out to be one of the growth areas of the first half of the 21st century. So the relationship between sustainability issues and the city's economy will not simply be a question of perceived opposites in conflict.

These issues will not go away. The events described in this chapter have changed not only the political agenda in Manchester but also the ways in which these issues get addressed, and it may well turn out to be that this is the most important legacy of the city's Local Agenda 21 process.

10

Some concluding remarks

Introduction

By its nature, this is a book from which it is not easy to draw general conclusions. Its primary purpose has been to show how the city planning process operates in practice, and if by this stage of the book some gains have been made in these terms then its purpose in being written has been served. In particular, I have tried to illustrate how the planning process in practice works through a series of dynamic interactions involving people, politics and policies as well as plans; hence the title of the book. The precise form that this takes may vary considerably between locations and over time, but there is probably a great deal of this that is in broad measure common to the planning process in a very wide range of situations.

Nevertheless, there is an obvious danger in generalising outwards from the experiences of a single city. I would assert that in some very important ways the major British cities are different from each other, and that indeed this is something that we should be celebrating. Uniformity is the father of caution, and therefore we should be very suspicious of attempts to make everywhere follow the same path or do the same things because this will stifle creativity and local initiative. One component of the future of our cities must be developing and then building on the asset base that they have, and their local strengths as places and as collecting points for people of drive and imagination must be part of this. We also know that what works in one area won't necessarily work in another, even if we are not always sure why this should be; but as long as it does have a chance of working somewhere, this should be enough to justify the process of finding ways to enable it to be tried without worrying too much about whether it will be a general panacea. These things all send out cautionary signals about the process of building a large edifice on localised foundations that might not bear the weight.

At the same time, there is a great deal in relation to our cities that they do have in common. Many of the economic and social trends that have buffeted them over the past couple of centuries have been broadly the same, although their local incidence has varied quite considerably. Government policy towards cities remains regrettably undifferentiated, and it is as yet unclear whether integrated government regional offices will see their prime function as being servants of their regions in Whitehall or servants of Whitehall in their regions. The powers and structures that local government has to enable it to make its contribution, and the obligations it finds itself under as well as (increasingly these days) the bidding processes in which it feels it has little choice other than to participate, are all part of an essentially top-down system of government that must be amongst the most centralised in the western world. These things suggest that the experiences of a single city may have some worthwhile questions to raise for us about more general matters.

This chapter therefore tries to draw together some of the key issues and challenges that have emerged throughout this book, both to try to emphasise these points in the specific Manchester context and to provide a basis for readers to consider the extent to which experiences from elsewhere might parallel these. It begins by reflecting on what the experiences recorded in this book may have to tell us about the customer-driven approach to the planning service that was argued for. It then proceeds to look again at the roles of development planning and development control as the core functions of the planning service. This leads in turn to an identification of four key issues that are likely to determine how successful Manchester will be in seeking to continue the process of regenerating itself beyond the millennium, which readers who have persevered thus far will readily recognise as having been picked up on several occasions in this book. It itemises a series of factors which will influence very considerably how successful the city actually is in pursuing these questions, assuming that there is the political will to do so. Finally, it looks at some research issues and questions that have been suggested by all this.

The customer-driven approach to the planning service

It should be clear from what has been said that there are two central tensions in the application of the customer-driven approach with which this book commenced:

1) The tension between what customers hope and expect to receive from the planning service and what it is capable of delivering.
2) The tension between a responsive attitude to the expressed wishes of customers and the place of the planning service as an element in the hierarchy of local government which is ultimately politically controlled.

These two central tensions are not likely to be capable of being wished away, but equally their continued existence does not invalidate the general argument in favour of a customer-driven approach. I hope I have said enough in this book to demonstrate beyond much doubt that the historical record simply does not support the view that the planner knows best, and also to show that for planners and planning to achieve things that improve the places for which they work and thereby for the people whose lives are affected by the ways those places function it is necessary to work with those people. Thus, the development of thinking about customer-driven approaches to service delivery needs to accommodate these two central tensions as an integral part of the task. It may well be that in an ever-more pluralist society, and with constant pressures for the work of local government to be ever-more open to public scrutiny, these trends will of themselves help with the resolution of difficulties created by the continued existence of these two central tensions. Whilst there aren't any simple prescriptions for dealing with these matters, there are some relatively simple concepts that are likely to be important components of this, even if their simplicity of expression may seem somewhat elusive when it actually comes to the process of trying to apply them in the real world.

These simple concepts are set down as the seven characteristics of planner behaviour described in the conclusions to Chapter 2. They can be boiled down to seven key words: *competence, integrity, fairness, listening, flexibility, opportunism* and *relationships*. These basic components of professionalism rely on the inculcation of skills and values in initial educational processes that are capable of being honed throughout a life of practice, constantly learning from the experiences that this provides; and it is hoped also, learning from other people's experiences if as a profession we can get much better at writing these up in accessible ways than we are at present. These things will not remove the problems of customer expectation or the limitations that arise from operating within a local government base that have been described as being the central tensions inherent in the customer-driven approach to planning, but they will make us better at recognising and at accommodating these tensions as part of the development of that approach. In particular, planning has to see itself as being part of the democratic process. It is no part of the arguments that have been presented in this book that planning should be seen as anything other than an activity carried out on behalf of the public at large, and this must mean that it is undertaken in public arenas, subject to proper public scrutiny, regulated by laws and conventions, and responsible ultimately to democratic control. There may be many failings with each and every step in this process, some of which have been recorded in this book, but none of these amount to an argument for seeing planning ultimately as anything other than a public good. Customer-driven approaches to the delivery of planning services must therefore start from this perspective; the customers *are* the public whose good this is.

Development planning and development control revisited

Chapters 4 and 5 have said a great deal about how the development planning and the development control processes operate in practice. Chapter 4 has argued for some quite significant improvements that could be made to the development planning process as it is in Britain today, which would undoubtedly prove controversial in some quarters. Chapter 5, on the other hand, has argued for a more limited series of changes in respect of development control, and has instead suggested that the emphasis needs to move away from the speed of decision-making (without suggesting that slowness is of itself anything other than an undesirable characteristic) and towards a greater emphasis on the quality of the decisions that emerge. This movement needs to be reflected, amongst other things, in a range of nationally acceptable performance indicators for the local government planning service which reflect the quality and the extent of what it does rather than (as at present) the exclusive concern with speed of decision-making in development control. The central point, however, is that development plans and the ability to control development are amongst the key points of intervention that are available to planning in the process of regenerating cities; and thus how effective it is in that function is likely to depend substantially upon how good these basic tools are in the first place, and then upon how well they are applied to real-world situations and to the process of relating to customers in attempting to find satisfactory solutions to the problems posed by those situations.

This is why the arguments about 'quick in: quick out' approaches to development planning are so important. Development plans can take such a long time to produce and then to take though the statutory system in Britain that there is a real risk that they can be at least in part out of date by the time they are adopted. The sheer effort involved in going through all this can also make local planning authorities understandably reluctant to begin to think about changing quickly that which has been recently hard won. And the advent of Section 54A of the Planning Act, whilst important in reasserting the significance of the development plan, is likely if anything to exacerbate these difficulties. I think the way forward in all this is to try to make the development plan system in Britain more 'bottom up' than it is at present. At the end of the day, the thing that is most important in any plan is its relevance to the situation on the ground. To meet this test we need to develop more localised approaches which recognise that there are likely to be families of different types of plans, to which different types of statutory procedures can apply; and one large step in this direction would be to move towards giving local planning authorities a general power of competence rather than by seeking (as at present) to prescribe or to regulate what almost appears to be their every move in development plan-making. The national components of the planning system ought to be about those matters of national and regional policy on which government genuinely ought to be expressing views, and clearly there need to be mechanisms which ensure that these matters don't get discounted at the local level. But for the most part these are not likely to be as numerous or as determining at the local development plan scale as are the series of local issues to be tackled and local aspirations to be addressed. A development plan-making system which is congruent with these approaches may well also be a comfortable base for a development control system which is focused on quality of decision rather than on speed. After all, in 10 or 20 years time the speed of a bad decision will be irrelevant to the fact that the populace of an area may be stuck with the consequences of that bad decision for at least another 10 or 20 years, and often longer than this.

Development plans and development control can be very powerful tools in the regeneration process if they are harnessed to its other components. They give planners a very considerable potential to play a constructive role in that process, but they will only be able to do this if these tools are seen by the other participants in the process as being valuable in these same terms. Merely relying on the statutory force that goes with this territory is never likely to be enough, and it can all too easily be seen as defensiveness or even as negativity. What we therefore need to do is to improve the utility of our tools by improving both their quality and their appropriateness; and Chapters 4 and 5 have advanced some suggestions as to how this could be done.

Interestingly, there appears to be an element of thinking along the lines of giving primacy to the issue of local appropriateness in the recent work of the Local Government Commission for England. The commission's draft recommendations for the north Kent area (Local Government Commission for England, 1995), for example, focus quite extensively on the nature of the strategic planning job to be done in an area which the government has already decided is to experience major urban growth in the following decades in the form of the Thames Gateway initiative. The commission's conclusions (pp. 50–3) are that one of the proposed two new neighbouring unitary authorities should be instructed to prepare a unitary development plan of the type found in the metropolitan areas, and the other should become a structure planning authority on a joint basis with Kent County Council, thus carrying forward existing strategic planning arrangements into the new situation. This recommendation is based upon an analysis of the planning circumstances of each of the two new authorities, and an assessment of the appropriateness to these circumstances of the plan-making arrangements currently available. Whatever the local arguments that there will undoubtedly be about this, and I know nothing about the particular circumstances of north Kent, the interesting feature of this analysis is that consistency is seen to be of much less importance than is appropriateness. If the Local Government Commission can adduce this principle for a situation of structural change, isn't there also a strong argument for saying that it could also apply in situations where structural change is not necessarily contemplated? At the time of writing, these are only draft recommendations on the part of the commission and it will be interesting to see how these arguments about appropriateness will fare when confronted in the political world with arguments about consistency and conformity, but the point about a willingness to encourage variety in the system will not go away whatever this outcome.

Four key issues for Manchester

The material presented in Part II of this book was selected to demonstrate my view that there are four issues of over-riding importance that Manchester will have to tackle in seeking to continue the process of regeneration:

- The need to secure the economic base of the city.
- The need to ensure that the 20–30 per cent of the population of cities like Manchester that are seriously economically and socially deprived participate both in the city's wealth and in its governance.
- The need to create an urban transportation package that works in the long-term interests of the city.
- The need to move towards urban sustainability.

Overall, I would argue that Manchester's track-record shows that it has done well on the first of these and has been making some good progress with the fourth over a relatively short period of time, albeit with some local political difficulties, but has done much less well on the second and the third. One of the fairly clear things about this analysis is that it more or less reflects the differences between fields in which a great deal can be done in the locality and fields which are very dependent upon wider government policy and financial support; although of course these are differences of degree rather than of kind. The council got on with the things it was in a position to get on with, and was able to demonstrate some positive achievements as a result, but the things that couldn't be tackled effectively by local action but either were constrained

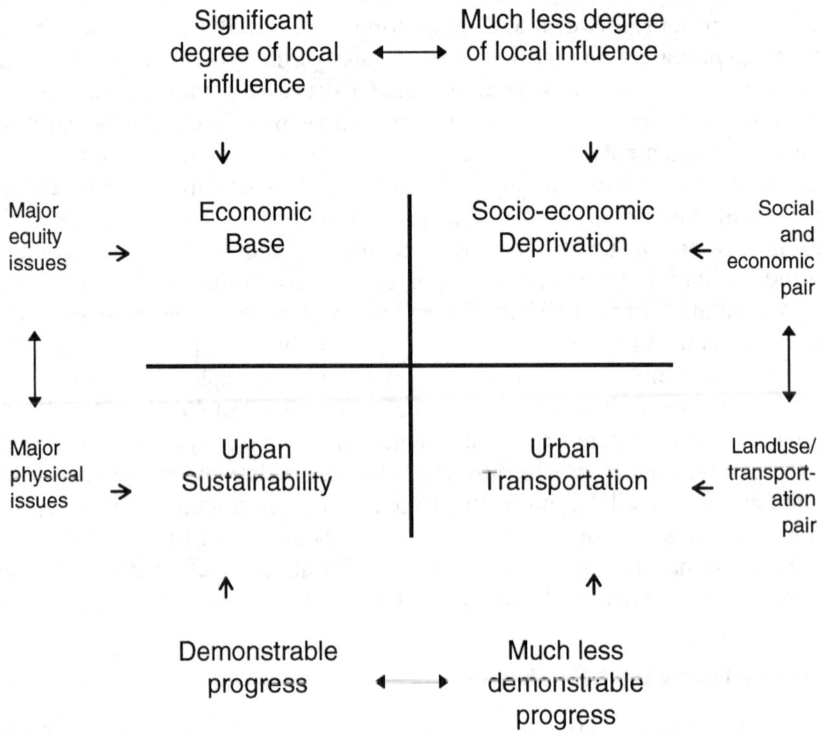

Figure 10.1 Key issues on the emerging urban agenda

by national policies or needed major and continuing national financial resource support had seen little progress (and in some cases several backward steps) because these had not been forthcoming. No doubt a full analysis would show that the relationships here are much more complex than this, but I don't believe it would change this basic conclusion. Indeed, it would be hard to argue with the proposition that urban deprivation in many parts of Manchester appears to be getting worse; and it would be equally hard to argue that the current shambles that poses for urban transportation arrangements amounts to the kind of integrated system that our cities clearly need; and it would be impossible to argue that local authorities in contemporary Britain are able to go out and change these situations by themselves. Perhaps the key point about these issues, however, is their mutually reinforcing (or mutually undermining) nature. The process of tackling any one of them inevitably has implications for the process of tackling the others. What this suggests, therefore, unless it is going to be asserted that one is of over-riding importance, is that ways need to be found of moving forward on all four together.

Figure 10.1 shows these issues as a matrix, organised according to some of the key variables. The horizontal line in Figure 10.1 denotes what are probably the most natural pairings in this cluster, where one pairing can be regarded as being broadly socioeconomic in nature and therefore as raising major equity issues and the other can be regarded as being broadly about land-use/transportation matters and therefore as raising major physical issues. When looked at like this, each of these two clusters contains one field where some genuine progress has been made and one field where very little progress can properly be claimed in recent times. The vertical line on the diagram divides those fields where a considerable amount can be done by local action (which are on the left-hand side of the diagram) from those fields which are so heavily dependent upon national policies or national resource support that very little can be achieved in them by purely local action. I am arguing here that this distinction is an important explanation for the very mixed record of achievement in the natural pairs of issues described above.

The example of economic base work illustrates very well the point made above about the interactive nature of these four key issues. It is too easy to assume that economic base work will automatically benefit the most deprived sections of the community, but the evidence does not support the view that this will happen automatically though a form of trickle down. Manchester over the past few years has undoubtedly made strides in dealing with its economic base in the manner described in Chapter 6, but at the same time many of the circumstances of many of its inner-city residents that are described in Chapter 7 have actually deteriorated. The position not only in Manchester but in Greater Manchester in these terms is analysed in more detail in Robson *et al.* (1994, pp. 215–32 and 328 and 329), and that analysis supports this broad conclusion. This is not, of course, an argument against doing work to improve the city's economic base; this must go on all the time. But it is an argument that says that this work must be paralleled by work that addresses the real economic and social circumstances of its most deprived citizens, and we must not assume that this will follow automatically from high-profile work on the city's economic base. We probably need to look much harder at how cross-overs from the one to the other can be achieved, because we must not give up on this as a possibility until demonstrably it isn't. I also suspect in practice that this will mean a lot more attention to local circumstances and to working with local people on finding ways of translating their ideas into reality than we have yet managed to achieve, and the development of effective partnerships at the very local level may have a great deal to contribute to this process (Short, 1989; Atkinson, 1995). These broad arguments are also canvassed in more academic terms by Lovering (in Healey *et al.*, 1995, pp. 109–26).

The same point also applies to the relationships between economic base work and the other two key issues, and has been illustrated particularly in the tensions over the approaches to the restraint of the private car in the city centre discussed in Chapters 8 and 9. A more radical approach in Manchester to private car restraint in the city centre would undoubtedly have a role to play in the development of sustainability policies, and may (although this is clearly more arguable) also have a role to play in encouraging greater use of public transport. But it was not pursued as a policy because of fears about its impact on the functioning of the city centre's economy. Whilst it can certainly be argued that experience from elsewhere does not demonstrate that these fears would be bound to be realised (Hass-Klau, 1993), it also needs to be acknowledged that they are understandable because one can never know what the outcome of a particular policy prescription is in a particular location until it has been tried. I would say that the lessons to be drawn from this are a combination of experiment and measure; try something on a small scale first and test whether its outcome is as has been predicted or feared, before stepping towards larger policy measures. What we have actually seen in Manchester, however, is an unwillingness to do this because the arguments about protecting the economic base of the city centre were seen by the council's political leadership to be of such over-riding importance as not to permit even the possibility of experiments which might carry risks of this kind, however small.

The need is to find ways of tackling these four issues together in mutually reinforcing ways. As has been said above, and as has been illustrated in several of the chapters of this book which deal with individual topics, this cannot all be done at the local level. Government will need to find more effective ways of maintaining a long-term commitment to the regeneration of the inner city and to improving the quality of life of its citizens if the task of reducing the scale of social and economic deprivation is to be accomplished. And government needs to find a set of transport policies that can work effectively in our cities rather than pursue policies based largely on dogma, that focus primarily on ownership and on the belief in competition however artificial this is rather than on the quality of the service available to customers. These will have to be accompanied by a series of developments at the local level, however; just as it is clear that the regeneration of the inner cities cannot be accomplished merely by action at the local level, so also is it the case that changes of policy and of attitude at the national level will not achieve this objective by themselves. What is needed is for the top-down and the bottom-up approaches to meet in a sustained commitment to what will inevitably be long-term local partnerships, which focus on what can be made to work within particular localities with the close involvement of their people. It may then be possible to redraw Figure 10.1 in a less polarised form.

Although Figure 10.1 has been drawn to try to summarise experiences from Manchester in a very simple form, it may also be the case that it reflects more broadly the situation in respect of the urban regeneration agenda that may be found in many other British cities. The dangers of trying to generalise from the experiences of a single city have already been discussed, and it is clear that there will always be very important differences reflecting each unique mix of circumstances when looking at how the same broad set of phenomena materialise in particular localities. That having been said, I have tried this same broad analysis out several times in discussions with colleagues in other cities, and the reaction has tended to be that in large measure it reflects their circumstances as well. Differences have tended to be about how well or how badly a particular area is doing in relation to one or more of these key issues, and about matters of policy emphasis; for example, the condition of the local housing stock may be a much more important issue in one locality than it is in another. If it is broadly true, however, that these four key issues and the interactions between them are central to the urban regeneration agenda today in the majority of our great cities, then the process of addressing this 'emerging urban agenda' (Kitchen, 1996b) is about much more than tackling more effectively the problems to be found in Manchester. This does not necessarily mean, of course, that the same package of solutions will or should apply everywhere, but identifying the most important issues is a good start towards generating an improved policy framework within which local initiative can be encouraged to flourish.

Some factors that will influence how these issues are tackled

There are four factors in particular that have emerged from the arguments presented in this book that are likely to play a very significant part in determining how effectively these four key issues are tackled:

1) The approach that is adopted to centralisation as against decentralisation, which was illustrated both in the context of the debate about planning in the inner city developed in Chapter 7 and in the context of the material on relationships with the political process presented in Chapter 3.
2) The development of partnership approaches, which have been discussed on several occasions throughout this book.

3) The development of interauthority and interagency co-operation, which was first touched upon in Chapter 2 when looking at the customers of the planning service and which surfaced subsequently on several occasions.
4) The whole question of the images of cities and of regions and of how those images are promoted, which was raised in particular in Chapter 6 on the economy of the city.

Each of these is taken in turn below.

The most basic difficulties with *centralisation* are that it seeks to replace

- diversity with uniformity; and
- a large number of potentially very wide-ranging views with a much smaller number of essentially conformist views.

These attempts don't always succeed in the long term, of course, but attempts to replace the many by the few whilst at the same time making life much simpler for those in control do also screen out a great deal of potentially useful views and ideas. This can happen both within central government and at the local level. The central government approach has tended recently to be one which achieves centralisation without increasing the apparent size of the central bureaucracy, through the creation of quangos (quasi-autonomous non-governmental organisations) with board members appointed by the Secretary of State. This is not merely a recent phenomenon either; it is perhaps salutary to remember that Chris Hood was writing about 'the rise and rise of the British quango' as far back as 1973 (Hood, 1973), and the pace of this and of other types of centralisation if anything increased during the years of Thatcher governments (Brindley, Rydin and Stoker, 1989; Thornley, 1991; Healey, 1992b). These sorts of things can happen at local level as well. Several chapters of this book have already talked about the growing centralisation of Manchester City Council's operations during Graham Stringer's leadership, and even the apparent flowering of various types of public–private partnerships in the city during this period may be somewhat of an illusion given the numbers of essentially the same 'local great and good' faces that get involved in many of these (Peck and Tickell, 1995, quote the phrase 'the Manchester Mafia' as a self-description by some of these people they interviewed). Perhaps more than anything else, what all this tends to exclude is the communities that are supposed to be on the receiving end of the jobs being done by this panoply of agencies and initiatives (Hill, 1994). Leaving aside the fact that they can at times be very difficult to deal with (as the Hulme experience demonstrated), it is hard to think of any arguments that would even begin to justify this. For all its messiness, greater decentralisation and a wider range of involvement must have a great deal to offer, and it is noticeable that the need to try harder to find ways of achieving greater local community involvement in the process of urban regeneration was one of the conclusions of the Department of the Environment's appraisal of the impact of British urban policy (Robson *et al.*, 1994). Indeed, it is clear that the broad conclusions

Box 10.1 The conclusions of the Department of the Environment's 1994 review of urban policy

My summary of the broad conclusions reached in the Department of the Environment's appraisal of the impact of British urban policy (Robson *et al.*, 1994), which was based primarily on an examination by three separate research teams of what had been happening over a period of time in the Merseyside, Greater Manchester, and Tyne and Wear conurbations, is as follows:

- Local collaborative partnerships need to be developed which are intended to operate in the long term.
- Local authorities have a significant role to play in improving the living circumstances of the people of their areas.
- Roles and opportunities need to be found to enable local communities to play a fuller part in these processes, and where there is a real problem of the lack of capacity to do this in local communities effort needs to be put into community capacity-building.
- The coherence of government programmes needs to be improved in terms of both their strategic objectives and the focusing of programmes and initiatives.
- Resources need to be targeted on the major urban areas in ways that remove much of the ambiguity that surrounds these processes at present, as to both the certainty of resource availability over time and also the consistency of policy applications. To this end, it was suggested that consideration should be given to the establishment of an urban budget to be administered at regional level.

of that study are very similar to the major themes that have emerged throughout this book, as a comparison of the contents of Box 10.1 with the contents of this chapter will readily demonstrate.

The point I want to make about *partnerships* is not that they are the magic ingredient that makes everything else work, but rather that we need to think much harder about the types of partnership that really can work and the circumstances in which this is likely to happen. In recent years, 'partnership' as a concept has appeared in Britain to achieve almost icon status, and yet the record here is much patchier than such a status would warrant and the form and purposes of partnerships much more diverse than is generally realised (Peck and Tickell, 1994; Bailey, 1995). It is clear from the work that was done to seek expert views as part of the Department of the Environment's appraisal of the impact of British urban policy (Robson *et al.*, 1994) that partnership was not seen as a universal panacea by consultees but as a series of ways of handling things with both strengths and weaknesses. This very much accords with my own experiences of working in and with various partnerships, which would suggest that they work best when they operate on the basis of mutual enlightened self-interest and least well when they are stretched to do things that go beyond the sorts of roles that their participants are comfortable with (Kitchen, 1996b). It is also worth saying that it is important to look at who isn't involved in a partnership and what it isn't talking about, in order to understand what it is doing and how it is doing it. Partnership is an exclusive as well as an inclusive concept. If the point about partnership working on the basis of mutual enlightened self-interest is right, then the starting point for understanding what it does must be the interests of its participants. This in turn links back to the point made above about the exclusion from or the marginalisation of many community representatives in partnerships.

Perhaps the sorts of partnership arrangements which might be particularly open to these sorts of criticisms are very tightly structured partnerships with very specific remits. I am very attracted by the potential of looser, much more open-ended and even more anarchic partnerships of the kinds that the City Pride process (see Chapter 7; see also Hall, Mawson and Nicholson, 1995; Newman, 1995; Williams, 1995a; 1995b; 1995c) and the Local Agenda 21 process (see Chapter 9) in their very different ways have been starting to show in Manchester. The point about these is that they are gatherings in various formats of quite large numbers of people who know something about particular areas of work or fields of activity, who are interested in an opportunity to take this further, and who also want to make a more general contribution to the development of the city through these sorts of processes. They too can be criticised for the fact that they don't always have a strong community input to them, although the Local Agenda 21 process is beginning to tackle this, and also because they have so far focused mainly on 'feel good' activities rather than on conflict resolution. It may be, indeed, that conflict resolution will turn out to require different kinds of mechanisms, although this is not a reason to ignore the positive potential of the kinds of partnerships being argued for here. What these sorts of mechanisms do offer, however, is a framework within which various kinds of *cognoscenti* do begin to talk more openly to each other and do begin to contribute their experience and understanding, and this must be a great deal better than an existing situation where too often people of this kind do not communicate with each other and do not as a consequence make the kind of contribution to the future of the city that they could. For this kind of mechanism to work effectively, however, there would need to be a greater openness in many political quarters to the diversity of ideas and opinions that it might produce. Diversity in these terms is a strength, but it may not produce the cosiness that is quite often a feature of much more limited partnerships between people who know each other already from several such other occasions and who often start off (perhaps as a result of those experiences) from essentially similar viewpoints anyway (Peck and Tickell, 1995). This argument for a 'partnership of the knowledgeable' is not meant as another form of exclusion, but is merely a way of saying that there are a lot of people out there who could contribute a lot more to the future of our cities and who would want the opportunity to do this; and we need to find more effective ways of enabling this to happen and then of coping creatively with its results.

If we could find more effective ways of creating 'partnerships of the knowledgeable', and then of linking these with community interests that can bring to the table fine-grained local understandings, we would have potentially a series of powerful and dynamic forces capable of adding richness and diversity (as well also political discomfort) to our processes of urban governance. These arguments in many ways echo the conclusions of Healey *et al.* (1995, pp. 273–89).

The point about linking 'partnerships of the knowledgeable' with local community interests is particularly important, because in the last analysis the primary asset of our cities is their people. This has been true throughout their history (Briggs, 1982) and in different ways remains true today. In recent times, for example, the regeneration of Chinatown in the city centre and of the centre of Rusholme just south of the city centre, respectively through mainly Hong Kong/Chinese and East African/Asian money, has repeated previous phases from Manchester's history of how people have come to the city and made their mark on it. We need to find ways of involving people much more widely in what we are doing to regenerate our cities; not only because this is their right anyway since they are on the receiving end of such activities; not even because their spirit and their inventiveness are essential ingredients in the mix; but because at the end of the day none of this will really work unless we take our people along with us (Colenutt and Cutten, 1994). The concept of partnership needs to be rethought radically in these terms. Local authorities have a particular role to play in all this by virtue of the leadership they can exercise in their communities, but they too have to be open to the variety that all this will engender. They need to be encouraged, as do we all, to accept and to welcome the need to go 'beyond the stable state' (Schon, 1971). Control is not the name of the game, and the achievement of a particular agenda by a political leadership will not be as important as the flowering of local initiatives that a truly enabling local authority may be able to encourage.

Interauthority and interagency co-operation have been with us for a long time, and will continue to be with us provided that the process of co-operation is seen by the potential co-operators as being advantageous to them. The story of the North West Regional Association illustrates this very well. In a region with no tradition of internal co-operation, the new association delivered over a relatively short period of time a regional economic strategy (PIEDA, 1993), a regional environmental inventory (North West Regional Association, undated), and draft regional planning advice to the Secretary of State for the Environment (North West Regional Association, 1994) because it was in the best interests of the parties concerned to do this. To give but three reasons, the recognition of the need for these sorts of actions seems likely to increase as economic processes globalise, as the debate about sustainability sharpens, and as the processes of working within the European Union mature. It simply is not the case in my experience, as is often alleged, that local authorities will not co-operate with each other. What certainly is the case, however, is that they need to be able to see that it is in their own best interests to do this before they necessarily will do so. I don't think in these terms that a lot of other organisations are really all that different. Interauthority and interagency co-operation is not about a romantic attachment to somebody else's broader ideals but is about practical action to address common problems, and I think in future there is likely to be more of it in the planning world because more problems will be seen as transcending local planning authority boundaries. Nor will the solution to this problem be found in searching for the perfect set of local authority boundaries, which in my view does not exist; whatever local authority boundaries are decided upon and for whatever reasons, issues will transcend them. What might help in England, however, is the creation of a tier of regional or provincial government, provided that this is sufficiently remote from the work of local authorities so as not to promote too many territorial disputes, and provided also that it is accompanied by some real and genuine decentralisation of functions and activities by central government.

The need to look at *the imaging and promoting of cities* (and for that matter of regions) really follows on from this (Kearns and Philo, 1993). Manchester's Olympic bid experience has lessons to offer in these terms (Kitchen, 1996f). The example that is often quoted of an effective city campaign is that of Glasgow (Paddison, 1993), which appeared to achieve a great deal in turning round the image of that city and creating one that is in many ways nearer to the realities of a city that appeared to me to have far more going for it than it often appeared to realise itself. Both these examples show the importance of first believing in the message yourself. One obvious reason for this is that it is very difficult for others to accept a message from someone when it looks to them as if that person doesn't believe it him or herself. In many ways, the City Centre Campaign in Manchester in the early 1980s (see Chapter 6) taught us that lesson, because it was clear from the feedback research at that stage that the ordinary citizens of Manchester had a more positive view of their own city centre than the council itself did. Another reason why this is important is that in practice the residents and users of a city have a vast number of personal contacts with all sorts of people all over the world, and daily transmit messages about the city as part of that process that in all probability exceed many times over the number of connections that can be made even by the most expensive of promotion campaigns. If these messages are negative ones, a promotional campaign will be fighting an impossible battle. If these messages are positive ones, however, any campaign that is still thought to be necessary can both build on this base and reinforce these messages. Thus the first stage in the process of city imaging and promotion must be the need to win the 'hearts and minds' campaign at home. At the end of the day, however, it is very difficult to believe a message if the visible reality is very different. Promoting cans of baked beans may be about the promotion of essentially similar things, with at best (or at worst, depending upon your perspective) subtle differences between the sauces as the only thing that really sets the contents of the cans apart. Thus, you are not in too much danger of being caught out by your own hype. But cities are not like this. If residents know and visitors can spot that rhetoric and reality are far apart in city promotion, then it is not likely to work. So attention to the quality of the product must accompany city promotion.

That having been said, the need for cities and regions to have positive images of themselves and to promote these effectively is surely not in dispute, in a world where image promotion and marketing seem to be becoming ever more important. Manchester is in competition with Birmingham, Liverpool, Sheffield, Leeds and Newcastle in some ways, but it is also in competition with Milan, Turin, Barcelona, Lille and Stuttgart, and this could also be repeated in the wider world. This is not just about the images that cities want to promote of themselves. It is also about the images that already exist. Many of these images relate to major associations that cities have; Manchester is indelibly linked with Manchester United Football Club, for example, and the fact that the club is located outside the city's boundary really makes no difference at all to this very strong image. Indeed, local government boundaries are largely irrelevant to all this. I have received many overseas visitors, for example, whose perception was that Manchester was much nearer to being a large industrial city with a population of about 2½ millions (in other words, the population of the whole of Greater Manchester) than to it being the administrative city with a resident population of about 440,000 at the core of this conurbation that in fact it is; and I am sure that this had a lot to do with the city's fame in terms of the history of the Industrial Revolution. This makes one of the most important points in the field that remains to be made, in my opinion, which is that we have to turn city and regional promotion from the internally competitive process that it is too frequently at present to a process which is internally co-operative and competitive only in relation to the real external competition. The north west has been riven with difficulties of this kind, which have meant that the efforts both in respect of its constituent elements and on behalf of the region as a whole have been much less effective than they need to be. This is one area where local government co-operation needs to go further, but as I have said above this will only really happen when mutual enlightened self-interest demonstrates to enough people in key positions that there need to be changes.

Some research questions

To my mind, there are two broad research areas that emerge from all this:

- First, we need to focus more effectively on what works where and why, and equally on what doesn't work where and why. Given that we have now had virtually 50 years of statutory planning activity, there is precious little material available that analyses very systematically what city planning processes are actually achieving and not achieving. There are many different ways of doing this, and the reflective contributions of planning practitioners should be one of those elements. But overall, we don't know anything like as much as we should about how effective our planning actions in cities are proving, what their intended and unintended consequences are, who is gaining and who is losing as a result, and why these outcomes are as they are. And until we know much more about these sorts of things, our attempts to feed back the results of experience into both practice and into better training and continuous development of professionals will be of limited worth. In fairness, in recent years the DoE's planning research programme has been showing clear signs of moving in this direction, but the willingness to evaluate what is being done openly and honestly needs to be much more widespread.

- Secondly, there are many more specific things that have been touched on in this book about which we need to know a great deal more. For example, we need to know more about who is actually benefiting from work to improve the economic base of our cities, how to make sure as far as possible that this includes our most economically and socially deprived population, and how to work effectively at a more localised level to target actions at helping these groups of people (and at helping them to help themselves). Quite a lot of work with these sorts of objectives is actually going on in our cities all the time. Much less of it is systematically monitored and reviewed, however, and even less of it is reported on in accessible ways. Of course, the political imperative with policy initiatives is always that they must work. Reporting back that they haven't done so, and on what the reasons for this might be, is politically very difficult at least while the politicians who initiated the action are still around. Without being naive about the sensitivities of these sorts of things, which will always be with us, we need to try to move towards a situation where it is openly accepted that initiatives are undertaken without knowing whether they will work, and that some actions (probably far more than we would usually care to acknowledge) are by their nature experimental. The willingness to monitor, to review, to consult and to report on these sorts of matters needs to be built more into our work than it often is, so that we can learn from each other's experiences without reinventing the wheel.

This brings me back neatly to where I started in writing this book. In support of these research activities, we need to find far more ways of encouraging practitioners, and perhaps particularly those who have had extensive and senior experience, to write up what they have been doing in reflective and accessible ways so that other people have a chance to learn from these experiences and so that we can improve our understanding of what experiences are 'typical'. It may well be, for example, that the experiences of the city planning process in Manchester that have been presented in this book are wholly typical of those of other cities in some senses, and wholly atypical in others.

The point is that we do not know, because the material that would enable us to make judgements of this kind is not available. This is letting British planning down in at least two ways. The first is that we must be in danger of repeating mistakes in a profession that has not developed a tradition of writing up its experiences as one of the ways of passing on knowledge and understanding from one generation to the next. The second is that this opportunity to contribute to policy evaluation work is simply not being taken. This isn't the only way in which policy evaluation work should be carried out, of course, but it is certainly one of the ways in which contributions should be sought. In particular, it might well be useful as a means of helping to understand why particular policies might be deemed to have succeeded or to have failed, which may well at least in part be about factors or elements that are not amenable to external research on which practitioners in those areas are well able to comment. If this book has contributed anything in these terms, I will be well pleased.

Postscript

The point has been made already in this book that many of the stories it tells are either unfinished or are parts of wider continuing processes. This postscript merely tries to add a further note about some critical developments in the period between the effective end of the process of writing the book and March 1997 (i.e. just before publication) in order to add a slightly longer-term perspective to some of the stories.

- The need to speed up *the development plan process* (Chapter 4) was acknowledged by the Government with the launch of a consultation paper in January 1997, although its contents do not go as far as some of the ideas I have suggested.
- *The IRA bomb in Manchester City Centre* (Chapter 6) which exploded on 15th June 1996 reportedly damaged 49,000 square metres of retail floorspace and 57,000 square metres of office floorspace at the heart of the centre. The task of responding by attempting to build a better city centre to replace the floorspace that was destroyed or irretrievably damaged is likely to be quite protracted, but a design competition has been held, a new public-private partnership has emerged to oversee the process, and significant public and private sector financial resource commitments have been made. As part of this process, there has been some acknowledgement of the opportunity offered here to reduce *the environmental impact of road traffic in the city centre* (Chapter 9).
- *The City Pride Process* (Chapter 7) produced a vision that still stands, multi-sectoral commitment both to the vision and to the process of generating it, and a series of project proposals which are being pursued in various ways. But it is difficult to see it as having given rise subsequently to the flowering of the new kinds of partnerships in Manchester of the kind I had hoped for at the time, and indeed it could be argued that in practice not much has really changed as a result of City Pride. Nevertheless, the Government in November 1996 invited seven more cities to prepare and submit City Pride statements, again on the basis that no additional finance was available for implementation. This generated a local response (which might

have happened anyway irrespective of this announcement) to the effect that for competitive reasons the Manchester statement needed to be reviewed and updated, with its eastern area widened to include parts of Tameside.

- *The regeneration of Hulme* (Chapter 7) has proceeded apace, although it wasn't complete when City Challenge came to an end on 31st March 1997. The momentum will hopefully be continued under new arrangements which will link together work on the Hulme and Moss Side areas.
- The final version of *the City Development Guide* (Chapter 7) was published early in 1997, and to my mind is significantly more generalised than its predecessor versions. I think this makes it a more helpful document, since its balance between specificity and flexibility is much improved by only being specific where this is necessary. Indeed, looking at the version that has now materialised it is hard to see quite what all the fuss was about.
- Manchester Metropolitan University have began to implement their planning consent for a student hall of residence, which was at the heart of all the political problems illustrated by *the Stretford Road Case* (Chapter 7). The intention is that it will first accommodate students during the 1998/99 academic year.
- *Channel Tunnel rail link services from Manchester to London Waterloo* (Chapter 8) were eventually withdrawn before direct services to Paris and Brussels materialised, and this remains the situation. The worst fears of regional interests in the Section 40 passenger discussions have therefore come about, at least until through regional services begin to operate.
- Planning permission has been granted for a *second runway at Manchester Airport* (Chapter 8), and the proposed Metrolink extension to the Airport and to Wythenshawe has also been granted its basic powers. Environmental activists have begun to dig tunnels under and to occupy trees on the land affected by the second runway, and a large-scale environmental protest seems likely as the construction process gets under way.
- The draft of the *Manchester Local Agenda 21 Statement* (Chapter 9) was published early in 1997, and immediately ran into criticism from the City Council because it made some proposals in the civil aviation field (including fiscal measures to make air transport more expensive) which were seen as adversely affecting the interests of Manchester Airport and the direct and indirect jobs it creates or supports.
- Considerable potential for progress in the field of *city marketing* (Chapter 10) has been created through two public-private initiatives. The first was the establishment of Marketing Manchester, to try to provide a more coherent external image for the city (defined in its widest sense, and not merely by reference to its administrative boundaries) rather than to continue with a series of images each advanced via individual initiatives. The second is the creation of the Manchester Investment and Development Agency Service (MIDAS). Its aims are to attract new employment to an area roughly equivalent to the revised City Pride area discussed above, to

help existing companies to stay and to expand, and to tackle training needs so that local people will benefit from these job creation efforts. A particular hope is that MIDAS will continue the momentum established by Trafford Park Development Corporation, which is scheduled to disappear at the end of March 1998, in attracting new private sector investment to the vast Trafford Park industrial estate which has long been a major source of manufacturing and distribution jobs (primarily) for the Manchester conurbation. The two agencies are intended to operate in a complementary manner, one (MIDAS) with a specific focus on inward investment, job retention and development and training, and the other (Marketing Manchester) with a broad promotion and marketing function. These new arrangements are also intended to sit alongside the work of the regional inward investment agency (INWARD), although past relationships between INWARD and many of the Greater Manchester authorities have been difficult. It remains to be seen how this new set of relationships will work out.

References

Adams, D. 1994: *Urban Planning and the Development Process*. UCL Press, London.

Altshuler, A.A. 1969: *The City Planning Process: A Political Analysis*. Cornell University Press, Ithaca, NY.

Ambrose, P. 1994: *Urban Process and Power*. Routledge, London

Ankers, S., Kaiserman, D. and Shepley, C. 1979: *The Grotton Papers*. Royal Town Planning Institute, London.

Argyris, C. 1992: *On Organizational Learning*. Blackwell, Oxford.

Argyris, C. 1993: *Knowledge for Action*. Jossey-Bass, San Francisco, CA.

Association of Greater Manchester Authorities 1988: *Advice on Strategic Guidance*. AGMA, Manchester.

Association of Greater Manchester Authorities 1993: *Greater Manchester Economic Strategy and Operational Programme*. AGMA, Manchester.

Atkinson, R. 1995: *Cities of Pride: Rebuilding Community, Refocusing Government*. Cassell, London.

Atkinson, R. and Moon, G. 1994: *Urban Policy in Britain: The City, the State and the Market.* Macmillan, Basingstoke.

Audit Commission for Local Authorities in England and Wales 1989: *Urban Regeneration and Economic Development: The Local Government Dimensions*. HMSO, London.

Audit Commission for Local Authorities and the National Health Service in England and Wales 1992: *Building in Quality: A Study of Development Control*. HMSO, London.

Audit Commission for Local Authorities and the National Health Service in England and Wales 1994: *Manchester City Council. Profile 1993/94*. Audit Commission, London.

Audit Commission for Local Authorities and the National Health Service in England and Wales 1995: *Manchester City Council. Profile 1994/95*. Audit Commission, London.

Austin-Smith:Lord and JMP Consultants 1994: *Manchester Higher Education Precinct. Action Programme*. Jointly published by the three Manchester universities, the Central Manchester Healthcare Trust, Central Manchester Development Corporation and Manchester City Council, Manchester.

Bailey, N. 1995: *Partnership Agencies in British Urban Policy*. UCL Press, London.

Barlow, I.M. 1991: *Metropolitan Government.* Routledge, London.

Bartelmus, P. 1994: *Environment, Growth and Development: The Concepts and Strategies of Sustainability*. Routledge, London.

Bell, P. and Cloke, P. 1990: *Deregulation and Transport: Market Forces in the Modern World*. David Fulton Publishers, London.

Benveniste, G. 1989: *Mastering the Politics of Planning*. Jossey-Bass, San Francisco, CA.

Bianchini, F. and Parkinson, M. 1993: *Cultural Policy and Urban Regeneration: the West European Experience*. Manchester University Press, Manchester.

Blackman, T. 1995: *Urban Policy in Practice*. Routledge, London.

Blowers, A. 1980: *The Limits of Power: The Politics of Local Planning Policy* Pergamon, Oxford.

Blowers, A. 1993: *Planning for a Sustainable Environment*. Earthscan, London

Booth, P. 1996: *Controlling Development*. UCL Press, London

Bramley, G., Bartlett, W. and Lambert, C. 1995: *Planning: The Market and Private Housebuilding*. UCL Press, London.

Briggs, A. 1982: *Victorian Cities*. Pelican Books, London.

Brindley, T., Rydin Y. and Stoker, G. 1989: *Remaking Planning: The Politics of Urban Change in the Thatcher Years*. Unwin Hyman, London.

Brotchie, J., Batty, M., Blakely, E., Hall, P. and Newton, P. 1995: *Cities in Competition*. Longman, Melbourne.

Bruton, M.J., Crispin, G. and Fidler, P. 1982: Local plans: the role and status of the public local inquiry. *Journal of Planning and Environment Law*, May, 276–86.

Bruton, M.J. and Nicholson, D. 1987: *Local Planning in Practice*. Hutchinson, London.

Buchanan, C. 1963: *Traffic in Towns*. HMSO, London.

Burns, W. 1967: *Newcastle: A Study in Replanning at Newcastle upon Tyne*. Leonard Hill, London.

Caulfield, I. and Schultz, J. 1989: *Planning for Change: Strategic Planning in Local Government*. Longman, Harlow

Central Manchester Development Corporation 1995: *Housing Report Manchester*. CMDC, Manchester.

Cohen, R.A. 1977: Small town revitalization planning: case studies and a critique. *Journal of the American Institute of Planners*, Vol. 43, no. 1, 3–12.

Colenutt, B. and Cutten, A. 1994: Community empowerment in vogue or in vain? *Local Economy*, Vol. 9, no. 3, 236–50.

Committee of Vice-Chancellors and Principals 1994: *Town and Country Planning and University Estates*. CVCP, London.

Cross, D.T. and Bristow, M.R. 1983: *English Structure Planning*. Pion, London.

Cullingworth, J.B. and Nadin, V. 1994: *Town and Country Planning in Britain* (11th edition). Routledge, London.

Dalgleish, K., Lawless, P. and Vigar, G. 1994: *Urban Innovation and Employment Generation*. Office for Official Publications of the European Communities, Luxembourg.

Dalton, S.A. and Longhurst, J.W.S. 1994: *Towards a Sustainable Future: Promoting Sustainable Development*. Global Forum Academic Conference Advisory Committee, Manchester.

Davies, J.G. 1972: *The Evangelistic Bureaucrat*. Tavistock, London.

Deakin, N. and Edwards, J. 1993: *The Enterprise Culture and the Inner City*. Routledge, London.

Dennis, N. 1970: *People and Planning: The Sociology of Housing in Sunderland*. Faber & Faber, London.

Dennis, N. 1972: *Public Participation and Planners' Blight*. Faber & Faber, London.

Department of the Environment 1989: *Strategic Planning Guidance for Greater Manchester*. DoE, Manchester.

Department of the Environment 1992: *General Policy and Practice* (PPG1). HMSO, London.
Department of the Environment 1993a: *Development Plans: A Good Practice Guide*. HMSO, London.
Department of the Environment 1993b: *Enquiry into the Planning System in North Cornwall District by Audrey Lees*. HMSO, London.
Department of the Environment 1994a: *Partnerships in Practice*. HMSO, London.
Department of the Environment 1994b: *Quality in Town and Country*. DoE, London.
Department of the Environment 1995: *Air Quality – Meeting the Challenge*. HMSO, London.
Department of the Environment and Department of Transport 1994: *Planning Policy Guidance: Transport* (PPG13). HMSO, London.
Department of the Environment and Department of Transport 1995: *PPG13 – A Guide to Better Practice*. HMSO, London.
Duncan, S. and Goodwin, M. 1988: *The Local State and Uneven Development*. Polity Press, Cambridge.
Dunleavy, P. 1981: *The Politics of Mass Housing*. Oxford University Press, Oxford.
Dunleavy, P. 1991: *Democracy, Bureaucracy and Public Choice*. Harvester Wheatsheaf, New York.
ECOTEC and Transportation Planning Associates 1993: *Reducing Transport Emissions through Planning*. HMSO, London.
Elcock, H. 1994: *Local Government*. Routledge, London.
Elson, M.J. 1986: *Green Belts: Conflict Mediation in the Urban Fringe*. Heinemann, London.
Elson, M.J., Walker, S. and Macdonald, R. 1993: *The Effectiveness of Green Belts*. HMSO, London.
English Partnerships 1996: *Working with Our Partners*. English Partnerships, London.
Fenwick, J. 1995: *Managing Local Government*. Chapman & Hall, London.
Fischer, F. and Forrester, J. 1993: *The Argumentative Turn in Policy Analysis and Planning*. UCL Press, London.
Fudge, C. and Healey, P. 1984: *Local Planning in Practice (3): Camden and Manchester. University of Bristol School for Advanced Urban Studies Working Paper* 32. SAUS, Bristol.
Gans, H.J. 1972: *People and Plans: Essays on Urban Problems and Solutions*. Pelican, Harmondsworth.
Garside, P.L. and Hebbert, M. 1989: *British Regionalism 1900–2000*. Mansell, London.
Geddes, P. 1968 (first published 1915): *Cities in Evolution*. Ernest Benn, London.
Gilbert, R., Stevenson, D., Girardet, H. and Stren, R. 1996: *Making Cities Work: The Role of Local Authorities in the Urban Environment*. Earthscan, London.
Gladstone, F. 1976: *The Politics of Planning*. Temple Smith, London.
Glaser, B. and Strauss, A.L. 1968: *The Discovery of Grounded Theory*. Weidenfeld and Nicholson, London.
Goodman, R. 1972 *After the Planners*. Penguin, Harmondsworth.
Government Offices for the North West and for Merseyside 1995: *Regional Planning Guidance for the North West*. GONW and GOM, Manchester and Liverpool.
Grant, L. 1995: *Arena!* Cornerhouse Publications, Manchester.
Grant, L. 1996: *Built to Music: The Making of the Bridgewater Hall*. Manchester City Council, Manchester.
Gray, T.S. 1995: *UK Environmental Policy in the 1990s*. Macmillan, Basingstoke.

Greater Manchester County Council 1975: *County Structure Plan: Report of Survey: Employment and the Economy*. GMC, Manchester.
Greater Manchester County Council 1979: *Greater Manchester County Structure Plan: Written Statement*. GMC, Manchester.
Greater Manchester Visitor and Convention Bureau 1995: *1995 Audit of Tourism Business in Greater Manchester*. GMVCB, Manchester.
Greed, C. 1993: *Introducing Town Planning*. Longman, Harlow.
Greed, C. 1996: *Implementing Town Planning*. Longman, Harlow.
Gyford, J. 1984: *Local Politics in Britain*. Croom Helm, London.
Gyford, J., Leach, S. and Game, C. 1989: *The Changing Politics of Local Government*. Unwin Hyman, London.
Hague, C. and McCourt, A. 1974: Comprehensive planning, public participation and the public interest. *Urban Studies*, Vol. 11, no. 2, 143–55.
Hall, P. 1988: *Cities of Tomorrow*. Blackwell, Oxford.
Hall, P., Thomas, R., Gracey, H. and Drewett, R. 1973: *The Containment of Urban England* (2 volumes). George Allen & Unwin, London.
Hall, S., Mawson, J. and Nicholson, B. 1995: City Pride: the Birmingham experience. *Local Economy*, Vol. 10, no. 2, 108–16.
Hambleton, R. and Thomas, H. 1995: *Urban Planning Evaluation: Challenge and Change*. Paul Chapman, London.
Harrison, M.L. and Mordey, R. 1987: *Planning Control: Philosophies, Prospects and Practice*. Croom Helm, London.
Hass-Klau, C. 1993: *The Pedestrian and City Traffic*. Belhaven, London.
Haughton, G. and Hunter, C. 1994: *Sustainable Cities*. Jessica Kingsley and the Regional Studies Association, London.
Haywood, R. 1996: More flexible office location controls and public transport considerations. *Town Planning Review*, Vol 67, no. 1, 65–86.
Healey, P. 1983: *Local Plans in British Land Use Planning*. Pergamon, Oxford.
Healey, P. 1992a: A planner's day: knowledge and action in communicative practice. *Journal of the American Planning Association*, Vol. 58, no. 1, 9–20.
Healey, P. 1992b: The reorganisation of state and market in planning. *Urban Studies*,Vol. 29, no. 3/4, 411–34.
Healey, P. 1994: Development plans: new approaches to making frameworks for land use regulation. *European Planning Studies*, Vol. 2, no. 1, 39–57.
Healey, P., Cameron, S., Davoudi, S., Graham, S. and Madani-Pour, A. 1995: *Managing Cities: The New Urban Context*. Wiley, Chichester.
Healey, P., Davoudi, S., O'Toole, M., Tavsanoglu, S. and Usher, D. 1992: *Rebuilding the City: Property-Led Urban Regeneration*. Spon, London.
Healey, P., Doak, A., McNamara, P. and Elson, M. 1985a: *The Implementation of Planning Policies and the Role of Development Plans: Volume 1 – Main Findings*. Department of Town Planning, Oxford Polytechnic, Oxford.
Healey, P., Doak, A., McNamara, P. and Elson, M. 1985b: *The Implementation of Planning Policies and the Role of Development Plans: Volume 2 – Planning Policy Implementation in Greater Manchester and the West Midlands*. Department of Town Planning, Oxford Polytechnic, Oxford.
Healey, P., McNamara, P., Elson, M. and Doak, A. 1988: *Land Use Planning and the Mediation of Urban Change: The British Planning System in Practice*. Cambridge University Press, Cambridge.
Higgins, M., Prior, A., Boyack, S., Howard, T. and Krywko, J. 1995: *Planners as Managers: Shifting the Gaze*. Royal Town Planning Institute, London.

Hill, D. 1994: *Citizens and Cities: Urban Policy in the 1990s*. Harvester Wheatsheaf, Hemel Hempstead.

Hood, C. 1973: The rise and rise of the British quango. *New Society*, Vol. 25, no. 567, 386–88.

Howard, E. 1946 (first published 1902): *Garden Cities of Tomorrow*. Faber & Faber, London.

Hulme Regeneration Ltd 1994: *Rebuilding the City: A Guide to Development in Hulme*. Hulme Regeneration Ltd, Manchester.

Hutton, W. 1996: *The State We're In*. Vintage, London.

Imrie, R. and Thomas, H. 1993: *British Urban Policy and the Urban Development Corporations*. Paul Chapman: London.

Interdepartmental Review Team 1994: *Local Government Enforcement*. Department of Trade and Industry, London.

Jacobs, A. 1978: *Making City Planning Work*. Planner's Press, Washington, DC.

Jacobs, B.D. 1992: *Fractured Cities*. Routledge, London.

Jacobs, J. 1964: *The Death and Life of Great American Cities: The Failure of Town Planning*. Pelican, London.

Judd, D. and Parkinson, M. 1990: *Leadership and Urban Regeneration*. Sage, Newbury Park, CA.

Judge, D., Stoker, G. and Wolman, H. 1995: *Theories of Urban Politics*. Sage, London.

Kavanagh, D. 1971: *Economic Planning in the North West. Regional Studies Association Occasional Papers on Regional Plannning Organisation 4*. Regional Studies Association, London.

Kearns, G. and Philo, C. 1993: *The City as Cultural Capital, Past and Present*. Pergamon, Oxford.

Keith, M. and Rogers, A. 1991: *Hollow Promises? Rhetoric and Reality in the Inner City*. Mansell, London.

Kitchen, T. 1986: Inner city policy and practice 1975–1985: reflections on a lost opportunity. In Willis, K.G. editor, *Contemporary Issues in Town Planning*. Gower, Aldershot.

Kitchen, T. 1990: A client-based view of the planning service. *Planning Outlook*, Vol. 33, no. 1, 65–76 (also in Thomas and Healey, 1991 op. cit.).

Kitchen, T. 1993a: The Channel Tunnel and regional development issues in Britain. In Roberts, P., Struthers, T. and Sacks, J., editors, *Managing the Metropolis*. Avebury, Aldershot.

Kitchen, T. 1993b: Developing a quality public transport alternative. *Town and Country Planning*, Vol. 622, no. 3, 121–24.

Kitchen, T. 1993c: Olympic villages at home and abroad. *Housing and Planning Review*, Vol. 47, no. 6, 27–29.

Kitchen, T. 1993d: The Manchester Olympic bid and urban regeneration. In *Proceedings of the Town and Country Planning Summer School 1993*. Royal Town Planning Institute, London.

Kitchen, T. 1994: Sticking to the agenda. *Planning Week*, Vol. 2, no. 30, 14–15.

Kitchen, T. 1996a: A future for strategic planning – a Manchester perspective. In Tewdwr-Jones, M., editor, *British Planning Policy in Transition*. UCL Press, London.

Kitchen, T. 1996b: The Emerging Urban Agenda. *Department of Planning and Landscape, University of Manchester, Occasional Paper 43*. DPL, University of Manchester, Manchester.

Kitchen, T. 1996c: The heart of the north. *Town and Country Planning*, Vol. 65, no. 1, 7–11.

Kitchen, T. 1996d: Problems of policy adjustment in planning: the case of urban sustainability. In Blackhall, J.C. *Planning Education and the Profession*. Department of Town and Country Planning, University of Newcastle upon Tyne, Newcastle upon Tyne.

Kitchen, T. 1996e: The future of development plans: reflections on Manchester's experiences, 1945–1995. *Town Planning Review*, Vol. 67, no. 3, 331–53.

Kitchen, T. 1996f: Cities and the world events process, *Town and Country Planning*, Vol. 65, no. 11, 314–16.

Krumholz, N. and Forester, J. 1990: *Making Equity Planning Work*. Temple University Press, Philadelphia, PA.

Law, C.M. 1993: *Urban Tourism: Attracting Visitors to Large Cities*. Mansell, London.

Lawless, P. 1989: *Britain's Inner Cities*. Paul Chapman, London.

Lawless, P. 1996: The inner cities: towards a new agenda. *Town Planning Review*, Vol. 67, no. 1, 21–43.

Lawson, M. 1992: *Bloody Margaret: Three Political Fantasies*. Picador, London.

Lindblom, C.E. 1965: *The Intelligence of Democracy*. Free Press, New York.

Local Government Commission for England 1995: *Draft Recommendations on the Future Local Government of Dartford, Gillingham, Gravesham and Rochester upon Medway in the County of Kent*. HMSO, London.

Loftman, P. and Nevin, B. 1995: Prestige projects and urban regeneration in the 1980s and 1990s: a review of benefits and limitations. *Planning Practice and Research*, Vol. 10, nos. 3/4, 299–315.

Manchester Airport Company 1993: *Manchester Airport Development Strategy to 2005*. Manchester Airport Company, Manchester.

Manchester City Council 1961: *Manchester Development Plan. Written Statement*. Manchester City Council, Manchester.

Manchester City Council 1967: *Manchester City Centre Map 1967*. Manchester City Council, Manchester.

Manchester City Council 1984: *Manchester City Centre Local Plan*. Manchester City Council Planning Department, Manchester.

Manchester City Council 1992: *The Manchester Plan. The Unitary Development Plan for the City of Manchester. Deposit Draft*. Manchester City Council Planning Department, Manchester.

Manchester City Council 1994a: *Economic Development Statement 1994–95*. Manchester City Council Chief Executive's Department, Manchester.

Manchester City Council 1994b: *Sustainability in Manchester: A Strategy for Action*. Manchester City Council Planning Department, Manchester (Appendix 2 of this document reproduces the *UK Local Government Declaration on Sustainable Development* originally published in 1993 by the Local Government Management Board on behalf of the UK Local Authority Associations).

Manchester City Council 1994c: *City Pride: A Focus for the Future*. Manchester City Council Chief Executive's Department, Manchester.

Manchester City Council 1995a: *City of Manchester Community Care Plan 1995–1998*. Manchester City Council Social Services Department, Manchester.

Manchester City Council 1995b: *The Manchester Report: Outputs of Global Forum 1994*. Manchester City Council Planning Department, Manchester.

Manchester City Council 1995c: *City Development Guide: Draft: May 1995*. Manchester City Council, Manchester.

Manchester City Council 1995d: *Manchester: 50 Years of Change*. HMSO, London.

Manchester City Council 1995e: *The Manchester Plan. The Unitary Development Plan for the City of Manchester*. Manchester City Council Planning Department, Manchester.

Manchester City Council and Jarvis Management Training 1994: *Manchester Employment in Construction Charter*. Manchester City Council Chief Executive's Department, Manchester.

Manchester City Council Planning Department 1994: *Land Availability in Manchester*. Manchester City Council, Manchester.

Manchester Commonwealth Bid Committee 1995: *Manchester 2002: Commonwealth Games Bid*. Manchester Commonwealth Bid Committee, Manchester.

Manchester Evening News 1993: *Greater Manchester: 125 Years of Images from the Manchester Evening News*. Breedon Books, Derby.

Manchester Health for All Working Party 1993: *Health Inequalities and Manchester in the 1990s*. Manchester City Council, Manchester.

Manchester Olympic Bid Committee 1993: *The British Olympic Bid: Manchester 2000*. Manchester Olympic Bid Committee, Manchester.

Mansfield, N.W. 1970: Research into the Value of Time. *Department of the Environment Time Research Note* 16. Department of the Environment, London.

Marris, P. 1987: *Meaning and Action: Community Planning and Conceptions of Change*. Routledge & Kegan Paul, London.

Masser, I. 1983: *Evaluating Urban Planning Efforts: Approaches to Policy Analysis*. Gower, Aldershot.

McCallum, J.D. 1976: Comparative Study in Planning: Explorations of the 'Political Culture' of Planning in Britain and the United States. *University of Glasgow Discussion Papers in Planning* 7. Glasgow University, Glasgow.

McCarthy, P. and Harrison, T. 1995: *Attitudes to Town and Country Planning*. HMSO, London.

Midgley, D. 1994: *City of Manchester Unitary Development Plan: Report into Objections made to the Plan*. Manchester City Council, Manchester.

Morgan, P. and Nott, S. 1995: *Development Control: Law, Policy and Practice*. Butterworths, London.

Muchnick, D.M. 1970: *Urban Renewal in Liverpool*. Bell & Sons, London.

Newman, P. 1995: London Pride. *Local Economy*, Vol. 10, no. 2, 117–23.

Newton, K. 1976: *Second City Politics: Democratic Processes and Decision-Making in Birmingham*. Clarendon Press, Oxford.

Nicholas, R. 1945: *City of Manchester Plan, 1945*. Jarrold & Sons, Norwich.

North West Joint Planning Team 1974: *Strategic Plan for the North West*. HMSO, London.

North West Regional Association 1994: *Greener Growth*. NWRA, Manchester.

North West Regional Association (undated): *Environmental Action for North West England*. NWRA and North West Business Leadership Team, no publication location quoted.

Paddison, R. 1993: City marketing, image reconstruction and urban regeneration. *Urban Studies*, Vol. 30, no. 2, 339–50.

Paris, C. and Blackaby, B. 1979: *Not Much Improvement: Urban Renewal Policy in Birmingham*. Heinemann, London.

Peck, J. and Tickell, A. 1994: Too many partners . . . the future for regeneration partnerships. *Local Economy*, Vol. 9, no. 3, 251–65.

Peck, J. and Tickell, A. 1995: Business goes local: dissecting the business agenda in Manchester. *International Journal of Urban and Regional Research*, Vol. 19, no. 1, 55–77.

Pickup, L., Stokes, G., Meadowcroft, S., Goodwin, P., Tyson, B. and Kenny, F. 1991: *Bus Deregulation in the Metropolitan Areas*. Avebury, Aldershot.

PIEDA 1993: *Regional Economic Strategy for North West England*. North West Regional Association and North West Business Leadership Team, Manchester.

Planning Advisory Group 1965: *The Future of Development Plans*. HMSO, London.

Punter, J.V. 1990: *Design Control in Bristol, 1940–1990: The Impact of Planning on the Design of Office Development in the City Centre*. Redcliffe, Bristol.

Quarrie, J. 1992: *Earth Summit '92: the United Nations Conference on Environment and Development*. Regency Press, London.

Ratcliffe, J. and Stubbs, A. 1996: *Urban Planning and Real Estate Development*. UCL Press, London.

Ravetz, J. 1995: *Manchester 2020*. Town and Country Planning Association and CER Research and Consultancy, Manchester.

Reade, E. 1987: *British Town and Country Planning*. Open University Press, Milton Keynes.

Rein, M. 1983 *From Policy to Practice*. Macmillan, London.

Robson, B. 1988: *Those Inner Cities: Reconciling the Economic and Social Arms of Public Policy*. Clarendon Press, Oxford.

Robson, B. 1994: Urban policy at the crossroads. *Local Economy*, Vol. 9, no. 3, 216–23.

Robson, B., Bradford, M., Deas, I., Hall, E., Harrison, E., Parkinson, M., Evans, R., Garside, P. and Robinson, F. 1994: *Assessing the Impact of Urban Policy*. HMSO, London.

Robson, B., Bradford, M. and Tye R. 1995: A matrix of deprivation in English authorities, 1991 (Part 2 of Department of the Environment: *1991 Deprivation Index: a Review of Approaches and a Matrix of Results*). HMSO, London.

Robson, B., Topham, N., Deas, I. and Twomey, J. 1995: *The Economic and Social Impact of Greater Manchester's Universities*. University of Manchester, Manchester.

Royal Commission on Environmental Pollution 1995: *Transport and the Environment*. Oxford University Press, Oxford (first published by HMSO in 1994).

Rydin, Y. 1993: *The British Planning System: An Introduction*. Macmillan, Basingstoke.

Schon, D.A. 1971: *Beyond the Stable State*. Temple Smith, London.

Schon, D.A. 1983: *The Reflective Practitioner: How Professionals Think in Action*. Temple Smith, London.

Schubert, G.S. 1960: *The Public Interest*. Free Press of Glencoe, Glencoe, IL.

Short, J.R. 1989: *The Humane City*. Blackwell, Oxford.

Simmie, J.M. 1974: *Citizens in Conflict: The Sociology of Town Planning*. Hutchinson Educational, London.

Simmie, J.M. 1981: *Power, Property and Corporation*. Macmillan, London.

Skeffington, A. *et al.* 1969: *People and Planning: Report of the Committee on Public Participation in Planning*. HMSO, London.

Smith, R. and Wannop, U. 1985: *Strategic Planning in Action: The Impact of the Clyde Valley Regional Plan 1946–1982*. Gower, Aldershot.

Spawforth, P. and Rankin, R. 1995: Vox pop: local delivery. *Planning Week*, 6 July, 16–20.

Stein, J.M. 1995: *Classic Readings in Urban Planning*. McGraw-Hill, New York.

Stewart, J. 1983: *Local Government: The Conditions of Local Choice*. George Allen & Unwin, London.

Stewart, J. and Stoker, G. 1995: *Local Government in the 1990s*. Macmillan, London.

Stoker, G. and Young, S. 1993: *Cities in the 1990s*. Longman, London.

Taylor, I., Evans, K. and Frazer, P. 1996: *A Tale of Two Cities*. Routledge, London.

Tewdwr-Jones, M. 1995: Development control and the legitimacy of planning decisions. *Town Planning Review*, Vol. 66, no. 2, 163–81.

Thomas, H. 1994: *Values and Planning*. Avebury, Aldershot.

Thomas, H. and Healey, P. 1991: *Dilemmas of Plannning Practice*. Avebury Technical, Aldershot.

Thornley, A. 1991: *Urban Planning under Thatcherism*. Routledge, London.

Turner, G. 1967: *The North Country*. Eyre & Spottiswoode, London.

Turok, I. 1992: Property led urban regeneration: panacea or placebo? *Environment and Planning*, Vol 24, no. 3, 361–80.

Tye, R. and Williams, G. 1994: Urban regeneration and central–local government relations: the case of east Manchester. In *Progress in Planning*, Vol. 42, Part 1. Pergamon, Oxford.

United Kingdom Government 1988: *Manchester, Salford, Trafford: European Integrated Development Operation – a Programme of Action*. UK Government in conjunction with Manchester City Council, Salford City Council and Trafford Metropolitan Borough Council. No publication location stated.

United Kingdom Government 1990: *This Common Inheritance: Britain's Environmental Strategy* (Cm 1200). HMSO, London.

United Kingdom Government 1994: *Sustainable Development: The UK Strategy* (Cm 2426). HMSO, London.

Vandermeer, R., Mill, C.G. and Morrison, D. 1980: *Greater Manchester County Structure Plan: Examination in Public: Report of the Panel*. Department of the Environment, Manchester.

Vickers, G. 1983: *Human Systems are Different*. Harper & Row, London.

Wagner, F.W., Joder, T.E. and Mumphrey, A.J. 1995: *Urban Revitalization: Policies and Programs*. Sage, Thousand Oaks, CA.

Wannop, U. 1995: *The Regional Imperative: Regional Planning and Governance in Britain, Europe and the United States*. Jessica Kingsley and the Regional Studies Association, London.

Webman, J.A. 1982: *Revising the Industrial City: The Politics of Urban Renewal in Lyon and Birmingham*. Croom Helm, London.

Whitelegg, J. 1993: *Transport for a Sustainable Future: The Case for Europe*. Belhaven, London.

Whittaker, S. 1995: *First Steps: Local Agenda 21 in Practice*. HMSO, London.

Williams, C.C. and Haughton, G. 1994: *Perspectives Towards Sustainable Environmental Development*. Avebury, Aldershot.

Williams, G. 1983: *Inner City Policy: A Partnership with the Voluntary Sector?* NCVO Occasional Paper 3. Bedford Square Press, London.

Williams, G. 1995a: Local governance and urban prospects: the potential of City Pride. *Local Economy*, Vol, 10, no. 2, 100–108.

Williams, G. 1995b: Manchester City Pride – a focus for the future? *Local Economy*, Vol. 10, no. 2, 124–32.

Williams, G. 1995c: Prospecting for gold: Manchester's City Pride experience. *Planning Practice and Research*, Vol. 10, no. 3/4, 345–58.

Willis, K.G. 1986: *Contemporary Issues in Town Planning*. Gower, Aldershot.

Wilson, D. and Game, C. 1994: *Local Government in the United Kingdom*. Macmillan, Basingstoke.

Wilson, H. and Womersley, L. 1967 *Manchester Education Precinct*. Manchester City Council, Manchester.

Wolman, H. and Goldsmith, M. 1992: *Urban Politics and Policy: A Comparative Approach*. Blackwell, Oxford.

World Commission on Environment and Development (the Brundtland Commission) 1987: *Our Common Future*. Oxford University Press, Oxford.

Index